I0037843

"The aquapelago has become a seminal framework 1
island scholarship beyond the land-locked island and towards engagements
with watery surroundings. *Aquapelagos: Integrated Terrestrial and Marine
Assemblages* makes a major contribution to aquapelago thinking. It is not
only indispensable for island studies but significant for the shifting stakes
of broader debate in the social sciences, arts, and humanities. For, by foregrounding the power of thinking with terrestrial-aquatic continua, of emergent patchworks of dynamic relational (un)becoming, the aquapelago is a
powerful engagement with today's crisis of faith in modern frameworks of
reasoning, a unique challenge to the human/nature divide."

—**Jonathan Pugh**, *Professor of Island Studies,
Newcastle University (UK). Co-author (with David
Chandler) (2021) Anthropocene Islands: Entangled
Worlds, University of Westminster Press*

"What is an 'aquapelago'? The answer leads readers to question main-
stream understandings of socio-spatial existence. The opening discussion of
Hau'ofa's Pacific 'sea of islands' framework reveals 'the island' as a limited
construct stemming from a terrestrial, Euro-centric tradition. The chapters
then present examples by diverse authors who attend to the socio–political–
ecological–oceanographic dynamics of space and place in historical context,
combined with the critical stakes of global climate change. This is essential
reading for scholars emerging from generations of thwarted justice and
evaded settler colonial responsibility in search of reparative onto-epistemologies
to enable the sovereignty and survival of the world's most vital aquapelagic
assemblages."

—**Amelia Moore**, *Associate Professor
University of Rhode Island, USA*

"A dozen years after the concept was introduced to Island Studies, the long-
awaited international anthology dedicated to empirical and critical theoretical discussions on the aquapelago is here. It assembles a selection of essays
demonstrating the formidable scope and strength of the aquapelago as lens
for inquiry into the connection between land and sea and the position of
humans and non-humans in terrestrial and marine assemblages. This book
unlocks island scholarship's potential as analytic optic in progressive projects (re)thinking and (re)imagining environmental, social and existential
crises across the globe. Aquapelagos is a welcome collection that anyone
interested in shorelines – and in the consequences that changes in the ways
the water and the land influence each other – have on our lives."

—**Firouz Gaini**, *Professor, University of the Faroe Islands*

AQUAPELAGOS

Aquapelagos is a cross-disciplinary volume that is geared to a general undergraduate and non-specialist readership while also being rigorous and theoretically exciting for doctoral and advanced researchers of climate and ocean studies. It foregrounds marine-terrestrial assemblages as philosophical, navigational, and knowledge-making interfaces.

Drawing on ethnographic, geographic, architectural, sociological, and scientific methodies, *Aquapelagos* sheds light on varied approaches, dialogues, and responses to the catastrophic and impending futures unfolding across waterfronts from the Andaman Islands, Maldives, and Indonesia to the Grand Banks and the Juan Fernández Islands. It delves into pressing issues of human interrelations with aquatic environments, ocean volatility, ocean toxicity, flooding, inundation, mitigation, rising seas, and climate adaptation in interdisciplinary and comparative global terms. Within the conceptual framework of the aquapelago, the contributors to this volume explore aspects of integrated terrestrial and marine assemblages that enhance our understanding of the impact of global climate change and related rising sea levels on diverse planetary ecologies and the societies that depend on them.

The volume will be of interest to scholars, researchers, and students of ethnography, social anthropology, climate action, development studies, public policy, and climate change.

Philip Hayward is Adjunct Professor at the University of British Columbia, Canada, Editor of the journal *Shima*, and a Strategic Advisor for the River Cities Network. His research addresses oceanic, island, coastal, and riverine environments with particular regard to issues of cultural heritage, tourism, and representation. He has published articles in journals such as *Anthropocenes, Island Studies Journal, Lagoonscapes, Small States and Territories* and *Transformations,* and he has written and edited 14 books.

May Joseph is Professor of Social Science at Pratt Institute, Brooklyn, USA, and author of *Aquatopia: Climate Interventions* (2022); *Ghosts of Lumumba* (2020), *Sealog: Indian Ocean to New York* (2019); *Fluid New York: Cosmopolitan Urbanism and the Green Imagination* (2013); and *Nomadic Identities: The Performance of Citizenship* (1999). Joseph is co-editor (with Sudipta Sen) of *Terra Aqua: The Amphibious Lifeworlds of Coastal and Maritime South Asia* (2022); and co-editor of *Performing Hybridity* (1999). She co-edits three book series from Routledge: Critical Climate Studies, Ocean and Island Studies, and Kaleidoscope: Ethnography, Art, Architecture and Archaeology.

Critical Climate Studies

Managing Editors: May Joseph (US), Kavita Philip (Canada)
Commissioning Editors: El Glasberg (US), A. J. James (India)

The Critical Climate Studies book series is located in the transdisciplinary space that crosscuts the social sciences, humanities, creative writing, environmental studies, and climate science. Scholarship and activism are powerful but often invisible global forces, trapped in the interstices. We seek to draw attention and analysis to such domains. The series welcomes short books that experiment with holistic engagement, critique, and conversation about climate change, broadly conceived. In addition to nuanced academic prose from all disciplines, the series embraces multi-genre writing, experimental ethnographies, creative non-fiction, lyrical sociology, ficto-critical writing, as well as science-humanities collaborations. We encourage contributions that are investigative, immersive, and attentive to the understudied and obscured planetary transformations taking hold as climate change accelerates. Our interests lie in large debates as well as in the understudied regions and microhistories of the world, where the impact of the planet's climate convulsions generate altered experiences and analyses of ontologies, geographies, ecologies, and political economies.

Barbuda
Changing Times, Changing Tides
Edited by Sophia Perdikaris and Rebecca Boger

Aquatopia
Climate Interventions
May Joseph and Sofia Varino

For more information about this series, please visit: https://www.routledge.com/Critical-Climate-Studies/book-series/CCS

AQUAPELAGOS

Integrated Terrestrial and Marine Assemblages

Edited by Philip Hayward and May Joseph

Routledge
Taylor & Francis Group

LONDON AND NEW YORK

Designed cover image: Historic Twillingate fishing port, north-east Newfoundland (Getty Images)

First published 2025
by Routledge
4 Park Square, Milton Park, Abingdon, Oxon OX14 4RN

and by Routledge
605 Third Avenue, New York, NY 10158

Routledge is an imprint of the Taylor & Francis Group, an informa business

© 2025 selection and editorial matter, Philip Hayward and May Joseph; individual chapters, the contributors

The right of Philip Hayward and May Joseph to be identified as the authors of the editorial material, and of the authors for their individual chapters, has been asserted in accordance with sections 77 and 78 of the Copyright, Designs and Patents Act 1988.

All rights reserved. No part of this book may be reprinted or reproduced or utilised in any form or by any electronic, mechanical, or other means, now known or hereafter invented, including photocopying and recording, or in any information storage or retrieval system, without permission in writing from the publishers.

Trademark notice: Product or corporate names may be trademarks or registered trademarks, and are used only for identification and explanation without intent to infringe.

British Library Cataloguing-in-Publication Data
A catalogue record for this book is available from the British Library

Library of Congress Cataloging-in-Publication Data
A catalog record has been requested for this book

ISBN: 978-1-032-72344-0 (hbk)
ISBN: 978-1-032-94192-9 (pbk)
ISBN: 978-1-003-56953-4 (ebk)

DOI: 10.4324/9781003569534

Typeset in Sabon
by Deanta Global Publishing Services, Chennai, India

Dedicated to Alex Mesker, an inspirational colleague whose meticulous work on the journal Shima *over a twenty-year period provided the platform for many of the ideas raised in this volume.*

CONTENTS

FIGURES

CONTRIBUTORS

Natalia Gandara Chacana is lecturer and postdoctoral researcher in the Institute of History at the Pontificia Universidad Católica de Valparaíso (PUCV), Chile. She is also a member of the PUCV Climate Action Centre. Her research focuses on the history of science and knowledge about the Southeastern Pacific and the environmental history of Chile. Her latest research project studies the extinction of the Juan Fernández sandalwood.

Elizabeth Chant is assistant professor in Global Sustainable Development at the University of Warwick, UK. Her research interests include Latin American cultural studies, map history, travel writing, environmental humanities, and ephemera studies. She is currently developing a monograph on the trope of desolation in depictions of Patagonia. She has previously published on topics including the maritime charting of Patagonia in the 18th century and the short stories of Chilean author Francisco Coloane.

Arup Chatterjee is professor at OP Jindal Global University, India. He is the founding editor of *Coldnoon: International Journal of Travel Writing and Travelling Cultures* (which he ran from 2011 to 2018). He has authored a number of books, including *The Great Indian Railways* (2019), *Indians in London: From the Birth of the East India Company to Independent India* (2021), and *Adam's Bridge* (2024). He has been visiting scholar at Brunel University London and SOAS, University of London, and visiting professor at the University of the Faroe Islands.

Mike Evans is professor in Community, Culture, and Global Studies at the University of British Columbia (Okanagan). Trained first as an economic anthropologist working on issues of development in the small island nations of the Pacific (PhD McMaster 1996), he continues to participate in and contribute to the community of Island Studies scholars working in English. Now working in a participatory manner, and mostly with Indigenous communities in Canada, he has published across these fields in both traditional scholarly and community accessible formats.

Christian Fleury is associate researcher at ESO Caen, a social geography research center at the University of Caen, Normandy, France. He holds a PhD in geography on the topic of border islands in which he focused on Jersey, Saint Pierre, and Miquelon and Trinidad. His interests are centered on three areas of study: sea appropriation conflicts, island issues, and border effects, for which he has published a number of journal articles and book chapters.

Ayasha Guerin is assistant professor in the department of World Arts & Cultures/ Dance, University of California, Los Angeles. Guerin holds a PhD from New York University, American Studies and her first book project, *Making Zone A, Nature, Race and Resilience on New York's Most Vulnerable Shores*, focuses on the colonial foundations of the city's waterfront development from the 17th to the 19th centuries, tracing how conquest, slavery, and capitalism have physically altered coastal environments and ecological relations.

Philip Hayward is adjunct professor at the University of British Columbia, Canada, editor of the journal *Shima*, and a strategic advisor for the River Cities Network. His research addresses oceanic, island, coastal, and riverine environments with particular regard to issues of cultural heritage, tourism, and representation. He has published articles in journals such as *Anthropocenes, Island Studies Journal, Lagoonscapes, Small States and Territories,* and *Transformations* and has written and edited 14 books. https://www.island-researchph.com

Henry Johnson is professor in the School of Performing Arts, University of Otago, New Zealand. He has carried out fieldwork in various locations in Asia, Europe, and Australasia, with research publications covering the fields of ethnomusicology, Asian Studies, and Island Studies. His most recent books include *Handbook of Japanese Music in the Modern Era* (2024), *Music in the Making of Modern Japan* (2021), and *Nenes' Koza Dabasa* (2021).

May Joseph is professor of Social Science at Pratt Institute, Brooklyn, USA, and author of *Aquatopia: Climate Interventions* (2022); *Ghosts of Lumumba* (2020), *Sealog: Indian Ocean to New York* (2019); *Fluid New York: Cosmopolitan Urbanism and the Green Imagination* (2013); and *Nomadic Identities: The Performance of Citizenship* (1999). Joseph is co-editor (with Sudipta Sen) of *Terra Aqua: The Amphibious Lifeworlds of Coastal and Maritime South Asia* (2022), and co-editor of *Performing Hybridity* (1999). She co-edits three book series from Routledge, Critical Climate Studies, Ocean and Island Studies, and Kaleidoscope: Ethnography, Art, Architecture and Archaeology. www.mayjoseph.com

Jun'ichiro Suwa is associate professor in Cultural Anthropology and Oceanian Regional Studies at Hirosaki University, Japan. His research interests include music and performance in Oceania, Japanese folklore and cultural studies, music cultures in Romania and Central Asia, and anthropological theory in music. He is currently developing a monograph on the "water music" of Vanuatu women and a theoretical book in anthropology of music inspired by theorists such as Gilles Deleuze and Tim Ingold.

Jessica Vandenberg is postdoctoral researcher at Harvard University and a research fellow at the Ocean Nexus Center, USA. Drawing on political ecology, critical social sciences and multi-modal ethnographic methods, her research explores questions of power, knowledge, and equity related to ocean governance. Her research focuses on the rise of corporate environmental governance in marine restoration and remediation as well as paths towards decolonizing these spaces to prioritize diverse ways of knowing, reflexive and relational thinking, and nature–culture relations.

PREFACE

May Joseph

The Aquapelagic Turn

"The water will come" writes Jeff Goodell (2017). Taking this warning seriously, I organized a conference at the Pratt Institute in Brooklyn entitled 'Archipelagos and Aquapelagos: Conceptualising Islands and Marine Spaces' in March 2018, in collaboration with Philip Hayward, the editor of the journal *Shima* and conference keynote speaker. The event was intended to initiate a public discussion on the Eastern Seaboard of the United States about the New York archipelago and intercoastal structures across the United States. My impetus was to recognize the aquapelago of Lenapehoking, land of the unceded Leni Lenape, through an investigation of the amphibious histories of the New York archipelago (Joseph 2013). Gathered at the conference were a number of prominent Island Studies scholars from around the world, including Hayward, whose influential term "aquapelago" was the focus of interest for the weekend. The influential gathering of island scholars included Mike Evans, Christian Fleury, Ayasha Guerin, Carl Zimring and Jessica Vandenberg. The meeting laid the groundwork for this anthology of thinking about the idea of archipelagos, the importance of aquapelagic thinking within island structures and a situating of the New York archipelago within a wider conversation about low-lying small island states and related geologies in a global context.

Hayward's concept of the aquapelago allows an agile and interrelational way of thinking about islands, the borders between islands, shorelines and the connectivities across coastal and riverine landscapes that involve the interface of terrestrial and aquatic ecologies. As Hayward (2024a) explains, the concept of the aquapelago was introduced into Island Studies discourse in 2012 to refer to terrestrial and marine territories integrated by human livelihood activities. Drawing on Epeli Hau'ofa's articulation of the Pacific as a "sea of islands" (1993)", Hayward deepens the analysis of the junctures between "human and non-human elements within terrestrial and aquatic interzones" (2012a, 7) and identifies these accreted zones as assemblages, as performed entities within complex socioeconomic systems. Acknowledging the pioneering influence of Jane Bennett's *Vibrant Matter* (2010) on the concept of aquapelago as "living and inanimate entities that can be activated in various ways" (Hayward, 2024b, 2), he throws open the terrains of archipelagic coasts and island shorelines as rich sites of anthropogenic research. Aquapelagos are assemblages of dynamic aquatic systems interfacing marine relationalities (2012b). As Hayward's recent work on interspecies relations between human-seabirds and larger island ecosystems (2024b) suggests, the concept of the aquapelagic assemblages is a useful tool for conceptually exploring the full range of human and marine (and avian) biodiversity.

What is urgent about this collection of essays is the applicability of the term "aquapelago" to a wide range of climatological scenarios in the face of rising seas. "The world has always been an archipelagic entity with visions of the globe" writes the historian Dilip Menon (2020, xx). Once a marginal and esoteric area of inquiry within the growing field of Ocean Studies, a preoccupation with sinking islands and their agglomerations has brought to the fore the need for theoretical lenses with which to shine light on the myriad challenges rising across the shorelines of the world's populations, the majority of which dwell on islands between East Africa, South Asia and the South China Seas. Eric Tagliacozzo's *In Asian Waters: Oceanic Worlds from Yemen to Yokohama* (2024) persuasively demonstrates through a sweeping account of Asia's maritime history how the regions' shoreline networks wove an intricate web of connectivity shaping our planet today. Another pathbreaking recent study of the importance of an aquapelagic approach to the study of environmental histories, Carl A. Zimring and Steven H. Corey's *Coastal Metropolis* (2021), foregrounds the coastline of archipelago New York as a place whose coastal infrastructures of harbor, estuary, archipelagos and aquatic discardship requires rewriting through the history of wastescapes – which are aquapelagic structures of contamination.

Critical to these recent contributions to water thinking is the working principle that Hayward has delineated as the "aquapelago". The notion of the aquapelago directs critical inquiry onto the terraqueous boundaries between land and sea, and across the terrestrial and marine divide. The zones of interrelationality between species and hydrologies emerge. Tenuous historicities elided because of land-based infrastructural thinking are exposed, as mapped out in Ted Steinberg's landmark book *Gotham Unbound* (2014). Steinberg's intricate excavation of the interface of land and water structuring the ecology of archipelagic New York served to shift thinking on New York thought towards a more water-bound engagement with the region's coast and informed our thinking on New York's aquapelagic future (Hayward & Joseph, 2024).

The aquapelago reveals traces of activities and dynamic ecosytemic processes that have coexisted across marine networks in far more entangled matrixes than formerly understood. Deploying the lens of the aquapelago sharpens the interdependencies between coasts and islands, of living waterfronts and marine ecosystems. As Menon underscores, "Regardless of where one is located on the globe, the rising waters press on our consciousness, and we are becoming more aware ... This sense of an interconnected history brings together for the first time the intertwined fates of humans, animals, and other beings on the planet, sentient or otherwise" (Menon 2020, 2). Propelled by this intensifying sense of risk and precarity impacting coastal resilience and climate predictability, the scholars in this collection address the question of aquapelagic knowledge production.

Seeing the Shoreline

Fernand Braudel (2001) emphasized that the sea can be fully understood only if we view it in the long perspective of its geological history. The essays gathered here exercise such a seeing in long perspective of little studied islands surrounded by what Helen Rozwadowski (2018) calls the vast expanse of oceans. Taking on Braudel's dictum that "the sea has to be seen and seen again" (Braudel 2001, 3), Hayward's notion of the aquapelagic assemblage threads a methodology of precarious perspectives onto Braudel's call for a geologic history, and his concentration on the remote and overlooked geologies of the sea, such as the Grand Banks of Newfoundland, is a deep dive into seeing the foggy shorelines far away from Braudel's Mediterranean. Hayward systematically articulates an aquapelagic thickening of ocean, air, ship and ice to the notion of foggy waterlines. Probing the outer peripheries of aquapelagic geographies over the last two decades, Hayward has accrued a variety of case studies of assemblages that are highly precarious for humans to negotiate(e.g., Hayward 2024c, 2024d).

In conversation with Hayward's focus on the remote, albeit with a more anchored approach, Jun'ichiro Suwa (2024) points out that the Japanese concept of *shima* entails the impact of human engagement in the aquapelago, explicating that the nature of island seeing in *shimaguni* incorporates politically coded spaces as homeland or a nation. Christian Fleury and Henry Johnson (2024) further develop the notion of the sea as a politically loaded space. They examine how the aquapelago enables a rigorous discussion regarding the tensions between areas such as the Minquiers and Écréhous reefs and international maritime boundaries. Well beyond the purview of the ways of seeing the sea around the Mediterranean, Jessica

Vandenberg's essay on the Spermonde Archipelago of Indonesia demonstrates how Hayward's aquapelagic framework allows an integrated reading "between islands, island people, and oceans" (Vandenberg 2024, xx). For Vandenberg, the aquapelagic framework is crucial to recognizing local and regional history as well as inter-island network systems in an effort to reassemble the older relationalities between coral restoration and remote island communities. Delineating the different ways of seeing aquapelagic societies in a different hemisphere, Elizabeth Chant and Natalia Gándara Chacana (2024) probe the transoceanic relationships shaping the Juan Fernández Archipelago and its islanders. Discussion of the over-exploitation and over-consumption of species endemic to Juan Fernández provides cautionary tales of forces impacting small islands in their study. Taking the malleable framework of the aquapelago into uncharted waters of the Bay of Bengal, Arup Chatterjee (2024) shows why the concept of aquapelagic assemblage is a salient framework for inquiry. Chatterjee argues that despite seeing and seeing again, the meanings accrued around an aquapelago and an ocean remain explosive and divisive. Meticulously unravelling the threads of islandic identities and geo-historicity in the marine passage between India and Sri Lanka across the Palk Strait, Chatterjee disentangles entrenched political and geological conflicts. In a similar vein, the decolonial work of Ayasha Guerin (2024) anchors the aquapelagic inquiries posed in the anthology by offering a Braudelian counterpoint to read the geology of the sea. Guerin disrupts the horizontality of Braudel's vision by submerging perspective and historicizing what lies under the sea through an analysis of places where land meets the ship. The sea for Guerin demands an aquapelagic lens that dissolves and curves around historical processes of exclusion, to shed light on the submerged in the aquapelago. Submergence, Guerin states, is a shared condition of a fugitive state.

Precarious Aquapelagos

My interest in spearheading this book on the aquapelago was particularly driven by my encounter with the work of Adam Grydehøj, Philip Hayward and the Island Studies communities who made me aware of my family's history of being islanders from the barrier islands of the Kerala coast (Joseph & Varino 2023). The idea of the aquapelago highlighted the low-lying regions and near shore islands of my cultural roots whose islandness have been mostly erased. The Malabar coast is ecologically and geologically different from the rest of mainland India. Along the Kerala coast one immediately sees the challenges of a land-based thinking that imposes terrestrial infrastructure on what is largely wetlands, barrier islands, floating sandbanks and a shifting, morphing coastal geology that comprises islets strung along the length of the largest waterbody in India, the Vembanadu Lake (Barlow

2022; Joseph 2019; Sen & Joseph 2022). This region of India is distinctly different in its island ontologies, and its infrastructure needs are largely subsumed under land-based development projects. The lack of island-based thinking and an environmentally sustainable approach to developing the region is resulting in a catastrophic collapse of local ecologies and disappearing biodiversity.

In an effort to open up the conversation about coastal thinking and foreground the specific terraqueous challenges facing South Asia, Sudipta Sen and I deployed Hayward's framework of the aquapelago in *Terra Aqua: The Amphibious Lifeworlds of Coastal and Maritime South Asia*. Our goal was to offer marine perspectives on a submerging landscape that is more amphibious than *terra firma* in coastal South Asia (Sen & Joseph 2022) and our impetus was propelled by the decolonial work of Island Studies scholars, such as Nadarajah, Martinez, Su and Grydehøj (2022). Pamila Gupta's (2021) nuanced thinking on the aquapelagic Global South as well as Meg Samuelson and Charne Lavery's work on the Oceanic South have been huge influences (Samuelson & Lavery, 2019). Of particular inspiration has been Lindsay Bremner's work on the coral atoll aquapelago of the Maldives, which paved the way for a reconceptualizing of how oceanic processes impact islands and geologies particularly in fragile ecologies such as reefs. Bremner's point that "Aquapelagos ... are performed geographies, not fixed spatial entities" (Bremner 2016, 22) draws attention to the complexities of global capital accumulation and the changing patterns structuring social relations between humans and the sea, what she calls "performing the aquapelago" (Bremner 2016, 22). Drawing on the Maldives, Bremner demonstrates how "the ocean became more enmeshed in human affairs" (Bremner 2016, 22) and methodically lays the case for the claim that:

> New aquapelagic performances combining infrastructures, technologies and global imaginaries produced new species of islands, new kinds of oceans, new typologies of architecture and new kinds of humans, constantly reassembling the unstable continuum among geological, hydrological, human, animal and technological life according to the laws of value.
> *(Bremner 2016, 22)*

As Amitav Ghosh's reading of the submerging Sunderbans and Asia's centrality to the climate crisis (Ghosh 2016) argues, alongside Dipesh Chakrabarty's call for a planetary thinking that takes habitability seriously (Chakrabarty 2021, 83), an attenuated and fluid approach to the aquapelagic is urgently needed. The ontology of wetness (Steinberg & Peters, 2015) has foregrounded inundation, subsidence, amphibiousness and sinking processes as key sources for aquapelagic infrastructural rethinking. Brian Roberts draws attention to the resulting conceptual challenge "How to understand the human and interspecies relations that cut and flow through (above, below, within, and with various and fractal currents) the planet's aqueous and ter-raqueous spaces?" (Roberts 2021, 112), and Helen Rozadowski calls for a recognition of the oceans past: "We must transform our understanding of the sea to one bound with history and interconnected with humanity" (Rozadowski 2018, 227).

This volume of essays – assembled by the editors of two of the lead-ing Island Studies journals, Hayward (*Shima*) and Joseph (*Island Studies Journal*) – offers some buoys towards a deeper understanding of the world's shorelines through the application of the idea of the aquapelago, which affords a way forward through the mire of foggy sandbanks and darkened waters.

References

Barlow, Matt. 2022. Floating ground: Wetness, infrastructure, and envelopment in Kochi, India. *Shima* 16(1), 26–44.

Bennett, Jane. 2010. *Vibrant matter: a political ecology of things*. Durham: Duke University Press.

Braudel, Fernand. 2001. *Memory and the Mediterranean*. New York: Vintage Books.

Bremner, Lindsay. 2016. Thinking architecture with an Indian Ocean Aquapelago. *GeoHumanities* 2(2), 284–310.

Chakrabarty, Dipesh. 2021. *the climate of history in a planetary age*. Chicago: The University of Chicago Press.

Chant, Elizabeth and Chacana, Natalia Gándara. 2024. The Juan Fernandez Islands in transition. In Hayward, Philip and Joseph, May (Eds.) *Aquapelagos: Integrated Terrestrial and Marine Assemblages*. New Dehli and New York: Routledge, xx–xx.

Chatterjee, Arup. 2024. The Entangled Island: Katchatheevu and Indo-Lankan maritime relations. In Hayward, Philip and Joseph, May (Eds.) *Aquapelagos: Integrated Terrestrial and Marine Assemblages*. New Dehli and New York: Routledge, xx–xx.

Fleury, Christian and Johnson, Henry. 2024. Making aquapelagic place in Jersey: The Minquiers and Écréhous reefs. In Hayward, Philip and Joseph, May (Eds.) *Aquapelagos: Integrated Terrestrial and Marine Assemblages*. New Dehli and New York: Routledge, xx–xx.

Goodell, Jeff. 2017. *The Water Will Come: Rising Seas, Sinking Cities, and the Remaking of the Civilized World*. New York: Little, Brown and Company.

Ghosh, Amitav. 2016. *The Great Derangement: Climate Change and the Unthinkable*. Chicago: The University of Chicago Press.

Guerin, Ayasha. 2024. We, the submerged: (Non)Humans, race and aquapelagic relations – Notes from New York. In Hayward, Philip and Joseph, May (Eds.) *Aquapelagos: Integrated Terrestrial and Marine Assemblages*. New Dehli and New York: Routledge, xx–xx.

Gupta, Pamila. 2021. Ways of seeing wetness. *Wasafiri* 36(2), 37–47.

Hau'ofa, Epeli. 1993. Our sea of islands. In Eric Waddell, Vijay Naidu and Epeli Hau'ofa (Eds.) *A New Oceania: Rediscovering our Sea of Islands*. Suva: University of the South Pacific, 2–16.

Hayward, Philip. 2012a. Aquapelagos and aquapelagic assemblages: Towards an integrated study of island societies and marine environments. *Shima* 6(1), 1–11.

Hayward, Philip. 2012b. The constitution of assemblages and the aquapelagality of Haida Gwaii. *Shima* 6(2), 1–14.

Hayward, Philip. 2024a. Introduction: The aquapelago as an integrated assemblage. In Hayward, Philip and Joseph, May (Eds.) *Aquapelagos: Integrated Terrestrial and Marine Assemblages*. New Dehli and New York: Routledge, xx–xx.

Hayward, Philip. 2024b. An avian-aquapelagic heritage at "The Edge of the World:" Reflections on humans and seabirds on St Kilda and the arrival of HPAIV. *Lagoonscapes* 4(1). https://doi.org/10.30687/lgsp/2785-2709/2024/01/003

Hayward, Philip. 2024c. The precarious aquapelagic assemblage of the Grand Banks (Northwest Atlantic). In Hayward, Philip and Joseph, May (Eds.) *Aquapelagos: Integrated Terrestrial and Marine Assemblages*. New Dehli and New York: Routledge, xx–xx.

Hayward, Philip. 2024d. The flower garden banks and the parameters of aquapelagic sanctuary. In Hayward, Philip and Joseph, May (Eds.) *Aquapelagos: Integrated Terrestrial and Marine Assemblages*. New Dehli and New York: Routledge, xx–xx.

Hayward, Philip and Joseph, May. 2024. New York: Lenapehoking/New York: An estuarine aquapelago. In Hayward, Philip and Joseph, May (Eds.) *Aquapelagos: Integrated Terrestrial and Marine Assemblages*. New Dehli and New York: Routledge, xx–xx.

Joseph, May. 2013. *Fluid New York: Cosmopolitan Urbanism and the Green Imagination*. Durham: Duke University Press.

Joseph, May. 2019. *Sealog: Indian Ocean to New York*. London: Routledge.

Joseph, May and Sofia Varino. 2023. *Aquatopia: Climate Interventions*. Delhi: Routledge.

Menon, Dilip M. 2020. Walking on water: Globalization and history. *Global Perspectives* 1(1), 12176.

Nadarajah, Yaso, Martinez, Elena, Su, Ping and Grydehøj, Adam. 2022. Critical reflexivity and decolonial methodology in island studies: Interrogating the scholar within. *Island Studies Journal* 17(1), 3–25.

Peters, Kimberley and Philip Steinberg. 2015. A wet world: Rethinking place, territory, and time. *Society and Place*. http://societyandspace.org/2015/04/27/a-wet-world-rethinking-place-territory-and-time-kimberley-peters-and-philip-steinberg

Roberts, Brian R. 2021. *Borderwaters: Amid the Archipelagic States of America*. Durham: Duke University Press.

Rozwadowski, Helen M. 2018. *Vast Expanses: A History of the Oceans*. London: Reaktion Books.

Samuelson, Meg and Lavery, Charne. 2019. The Oceanic South. *English Language Notes* 57(1), 37–50.

Sen, Sudipta and May Joseph. 2022. *Terra Aqua: The Amphibious Lifeworlds of Coastal and Maritime South Asia*. Delhi: Routledge.

Steinberg, Ted. 2014. *Gotham Unbound: The Ecological History of Greater New York*. New York: Simon and Schuster.

Suwa, Juni'ichiro. 2024. *Shima, shimaguni* and aquapelagic assemblages. In Hayward, Philip and Joseph, May (Eds.) *Aquapelagos: Integrated Terrestrial and Marine Assemblages*. New Dehli and New York: Routledge, xx–xx.

Tagliacozzo, Eric. 2024. *In Asian Waters: Oceanic Worlds from Yemen to Yokohama*. New Jersey: Princeton University Press.

Vandenberg, Jessica. 2024. Colonial legacies and restoration futures: Examining the risks of dispossession from coral reef restoration in the Indonesian aquapelago. In Hayward, Philip and Joseph, May (Eds.) *Aquapelagos: Integrated Terrestrial and Marine Assemblages*. New Dehli and New York: Routledge, xx–xx.
Zimring, Carl A. and Steven J. Corey (Eds.) 2021. *Coastal Metropolis: Environmental Histories of Modern New York City*. Pittsburgh: University of Pittsburgh Press.

ACKNOWLEDGEMENTS

Philip Hayward would like to thank Rebecca Coyle for encouraging and facilitating his research on the various aquatic projects in 2006–2012 that laid the groundwork for this volume. Thanks also to his daughters Amelia and Rosa for their support and tolerance of his endless research projects as they studied at school and, later, at university. This volume would not have been possible without the inspiring collaborations he has had with researchers involved with the journal *Shima* and with SICRI – the Small Island Cultures Research Initiative. He would particularly like to acknowledge the support and intellectual generosity of Connie Zeanah Atkinson, Rose Boswell, David Cashman, Denis Crowdy, Helen Dawson, Mark Evans, Mike Evans, Steven Feld, Jon Fitzgerald, Christian Fleury, Waldo Garrido, Ayano Ginoza, Ayasha Guerin, Felicity Greenland, Adam Grydehøj, Otto Heim, Sun-Kee Hong, Ian Hutton, Henry Johnson, Ilan Kelman, Junko Konishi, Sueo Kuwahara, Danny Long, Ian Maxwell, Marea Mitchell, Alahna Michele Moore, Jacobus B. Mosse, Peter Narváez, Karl Neuenfeldt, Hiroshi Ogawa, Jonathan Pugh, Amanda Reichelt-Brushett, Arianne Reis, Stephen Royle, Jun'ichiro Suwa, Francesco Vallerani, and Francesco Visentin. Despite being resolutely sceptical about the usefulness of the concept of the aquapelago, Godfrey Baldacchino also provided valuable commentary on related topics. The *Rising Currents: Projects for New York's Waterfront* exhibition held at New York's Museum of Modern Art in 2011 and its catalogue were inspirational in many ways, and he acknowledges the creative vision of editor/curator Barry Bergdoll and the participants. Dialogue and collaboration with various island and coastal-based creative artists has also been crucial and he would like to acknowledge Mauro Anselmo Olivos Castillo, George

Christian, Don Christian-Reynolds, Kath King, Jim Payne, Seaman Dan, Mick Thompson, Julie Toliman, and Titus Tilly for their input.

Thanks to Aakash Chakrabarty at Routledge for his efficiency and informative feedback during the production process. Also, thanks to Melanie Pannack for her extended patience as he worked on this volume from the earliest phase of their relationship. Lastly, he extends his gratitude to his editorial collaborator on this volume, May Joseph, who persistently encouraged him to develop a book-length project on aquapelagos and, thereby, brought this volume into existence.

Various institutions funded and/or facilitated research travel related to his aquapelagic research – the Australian Research Council, Macquarie University, Southern Cross University, and the University of Technology Sydney (in Australia); the University of Barcelona (Spain); Kagoshima University and Kansai University (in Japan); the Pratt Institute in Brooklyn and the University of New Orleans (USA); the University of Hong Kong; Udine University (Italy); and the River Cities Network, based at Leiden University (Netherlands). Many thanks for their support.

The final version of the manuscript for this volume was completed in Tórshavn (Faroe Islands), thanks to Firouz Gaini for showing me his fascinating aquapelago and stimulating fresh lines of thinking.

May Joseph would like to thank Aakash Chakrabarty, editor at Routledge India and the Critical Climate Studies Editors Kavita Philip, Viju James, and El Glassberg for their support of this project. Thank you to Philip Hayward for inviting me to collaborate on this special project. *Aquapelagos* is the result of a long collaboration between Philip Hayward, Adam Grydehøj, and the publications *Shima* and *Island Studies Journal*. Thank you to Sofia Varino for their collaborations across many islands. Thanks are also due to the extensive community of performers in Harmattan Theater <www.mayjoseph.com> who collaborated with me along the aquapelagic borders of islands and shorelines of the world – the Cape of Good Hope, Kochi, Venice, Isla Tiberina, Far Rockaways, Rotterdam, Lisbon, Amsterdam. To Godfrey Baldacchino, Su Ping, Ilan Kelman, Anne Schiffer, Laurie Brinklow, and Elaine Stafford, thank you for your stewardship and support. To my ocean colleagues Sudipta Sen, Smriti Srinivas, Devika Shankar, David Ludden, Francis Bradley, David Erdman, Pamila Gupta, Charne Lavery, Lisa Bloom, Arup Chatterjee, Dilip Menon, Neelima Jeychandran, Brian McGrath, Mikki Stedler, Jennifer Telesca, Edwidge Tamalet, Duarte Santo, Pedro Pombo, Michael Taussig, and Zhivka Valiavicharska, your intellectual generosities keep my writing in sight. To my islander cousins from Bolghatty Island, Vypin Island, the Alleppey coast, and the Kollam Ashtamudi Lake of Kerala – deep thanks for all the wisdoms shared over the decades about the challenges of living with sinking islands and an encroaching sea. Deepest thanks to my daughter Celine who was parentless for two terrifying days in New York City during Hurricane Sandy, for our shared learning about what it means to live on an island desperately unprepared for rising seas. Finally, I owe my ocean research to my ninety-year-old mother who still lives by the sea where she was born in Kollam, India.

Chapter 4 is a revised and updated version of 'Extraordinarily hazardous,' originally published in *Coolabah 34* (2023).

Chapter 5 is a revised and updated version of 'The risk of dispossession in the aquapelago,' originally published in *Shima 14* (2) (2020).

Chapter 6 is a revised and updated version of 'Sanctuary Islands in a hostile matrix,' originally published in *Island Studies Journal 14* (2) (2019).

1

AQUAPELAGOS

An Ontology of Integrated Aquatic and Terrestrial Assemblages

Philip Hayward

Introduction

During the late 20th and early 21st centuries, Humanities and Social Science research on islands and archipelagos tended to fixate on the terrestrial masses of islands and ignored or minimized the watery surrounds that created them. This tendency was particularly marked in the interdisciplinary field of Island Studies.[1] A number of researchers who had recognized and begun to analyze the deep interconnection of terrestrial and aquatic spaces in some island, archipelagic and coastal areas identified this flaw in meetings, conferences and individual articles but our critiques gained little traction. This was particularly frustrating as the perceptive and much-cited Pacific scholar Epeli Hau'ofa had emphasized the aquatic dimension of Oceanic societies in his characterization of the Pacific as a "sea of islands" (1993, 5). In terms of the arguments advanced in this volume, another term he uses is just as arresting, that of Pacific communities as "ocean peoples" (ibid: 8). As he goes on to characterize:

> "Oceania" connotes a sea of islands with their inhabitants. The world of our ancestors was a large sea full of places to explore, to make their homes *in*, to breed generations of seafarers like themselves. People raised *in* this environment were *at home with* the sea. They played *in* it as soon as they could walk steadily, they worked *in* it, they fought on it.
>
> *(Hau'ofa 1993, 1, [my emphases])*

The repeated uses of the term "in" here are just as pertinent as the characterizations of Pacific Islanders' navigations *across* oceanic surfaces. Hau'ofa

DOI: 10.4324/9781003569534-1

offers a model in which the immersive marine spaces of the Pacific's "sea of islands" are as fundamental an element of Oceania as its terrestrial extrusions. Indeed, the less frequently quoted poetic conclusion to his article emphasizes that:

> Oceania is humanity rising from the depths of brine and regions of fire deeper still, Oceania is us. We are the sea, we are the ocean, we must wake up to this ancient truth and together use it to overturn all hegemonic views that aim ultimately to confine us again, physically and psychologically, in the tiny spaces which we have resisted accepting as our sole appointed place, and from which we have recently liberated ourselves.
>
> *(Hauofa 1993, 16)*

It is fitting, in these regards, that the following discussions (and, indeed, the whole project of this volume) arose from my encounters and considerations of various Pacific epistemologies, experiences, cultural expressions and legal and organizational initiatives in the region. My work in this chapter – and, indeed, in my publications around the topic more generally – attempt to characterize and, essentially, 'translate' sets of Oceanic perceptions and possibilities into the thought- and word-scapes of Western anthropology, cultural geography and the island studies nexus. What follows is a set of ruminations on encounters, on conceptual insights, orientations and disorientations that I attempt to do justice to, informed by recognition of settler responsibility for societies that Westerners have severely disrupted (Bell, Lythberg, Woods & Yukich 2021, Garrison 2019).

Similar themes to those raised by Hau'ofa were expressed in the closing statement at the International Conference on Small Islands and Coral Reefs held in August 2009 in Ambon, in eastern Indonesia. The event was organized with a primary aim of sharing knowledge, information and experiences about the management and study of coral reef ecosystems. Within this, its main focus was on the need to ensure sustainable small island development in balance with ecosystem health and social justice for island communities and the waters than enabled them. An additional objective was to ensure that action plans were established that could address the impact of climate change on small islands. During the conference – which was primarily attended by regional delegates – the nature of archipelagos and archipelagic planning arose as key factors in debates. Reflecting these elements, a statement was drafted and formally presented at the conclusion of the conference, Item 1 of which identified that: "archipelagic regions, consisting of small islands and extensive marine territories including coral reefs, have very specific policy, planning and development requirements" related to their particular assemblages that required recognition and action on. Attending this event as editor of the journal *Shima* – which was established in 2007 to explore various themes in island cultures – I became aware of how inadequately we had grasped broader issues across regions and across terrestrial-aquatic continua.

One major shortcoming has been Island Studies' lack of understanding of and engagement with, marine spaces as complex three-dimensional systems. In particular, the field has been oddly reticent in responding to the range of insights, tools and data provided by modern oceanography. In some ways this lack is understandable (and has been reciprocated by oceanographers), given that oceanography has traditionally addressed the physical, geological, chemical and biological aspects of oceans. While these are important and informative, there has been a disconnect between them and the more sociocultural concerns of Island Studies that have limited dialogue and synthesis. These aspects were highlighted in a recent paper written by 21 UK-based oceanographers that calls for the establishment of a new field of "socio-oceanography" that the authors promote as "an opportunity to integrate marine and social sciences" (Popova et al. 2023). The collectively authored paper recognized that "the inherent interconnectivity between the ocean and society ensures that nearly everything we do in the marine natural sciences has the potential to influence and, perhaps address, ongoing and future societal challenges" and calls for cross-disciplinary collaboration on this account. It is too soon to judge how effective the intervention will be in fostering cross-disciplinary alliances between the natural and social sciences (let alone the Blue Humanities), but the aspiration is a highly positive one.

The previously identified shortcomings in Island Studies were exacerbated in the early 2010s when a number of Island Studies scholars began to call for the development of a "sister" field of archipelago studies. Despite Elisabeth DeLoughrey's clearly articulated concept of *archipelagraphy* as "a historiography that considers chains of islands in fluctuating relationship to their surrounding seas, islands and continents" (2001, 23) being available as an articulation point, early attempts to create Archipelagic Studies were largely premised on existing – and essentially "dry" – Island Studies approaches.[2] Aware that a decisive intervention was needed, a group of colleagues involved in researching islands collaborated to produce a series of short articles in the Debates section of the journal *Shima* in 2012–2014 to explore the topic.[3] One of the tactics we used was the invention of a new word to crystallize the concept of integrated terrestrial and marine environments and to precipitate discussion about it within the Island Studies field. Tactical use of neologisms is a tried and tested technique. Paul Crutzen's coining of the term 'Anthropocene' to refer to human-induced planetary change, for instance, has proved very effective as a focus for both scientific inquiry and environmental activism. In another context, the introduction of the term 'Postmodernism' to describe a range of Late Modernist sociocultural phenomena spawned a veritable academic industry ion the 1980s and

1990s. While I'm not comparing our project to the aforementioned ones, we used a similar tactic. The term we invented to give clear conceptual focus was the *aquapelago*. As might be apparent, the neologism plays off the more established term *archipelago* – putting *aqua* as the first syllable and thereby emphasizing that aspect. There are etymological complications to our coinage,[4] but the term has succeeded in getting the concept it represents on the map and has been used and developed by authors from various disciplines in a range of ways over the last decade.[5]

The *aquapelago* may be generally defined as:

> an assemblage of terrestrial and aquatic elements produced by human activities.

And the social manifestation of an aquapelago can be identified as one in which:

> the aquatic spaces between and around a group of islands are utilized and navigated in a manner that is fundamentally interconnected with and essential to the social group's habitation of land and their senses of identity and belonging.

These definitions serve to emphasize that while the *arch*ipelago is a physical concept, the *aqua*pelago is not. Rather, the aquapelago is an entity constituted by human activity. Neither cartography nor Google Earth (etc.) can identify an *aqua*pelago on its own – analysis of human utilization of space is necessary. There are further complexities that can be teased out of the above in that some humans inhabit aquapelagos more aquapelagically than others, i.e. not all humans who live on islands or coasts within aquapelagos are as actively engaged with integrated aquatic and terrestrial spaces as others. There may even be individuals within aquapelagos who (through intent or circumstance) exist outside of the aquapelagic assemblage, or – at the very least – minimally interact with it. (The most obvious examples being reclusive interior dwellers on islands or coasts who can sustain livelihoods without access to or interaction with aquatic realms and resources.)

Importantly, the concept of the aquapelago is not premised on a surface model; it also encompasses the spatial depths of waters. It provides a framework for understanding the continuum of terrestrial and aquatic resources and human activity and the connections between "cultural landscapes" created by agriculture and habitation and the aquatic-surface "scapes" and underwater "scapes" created by aquaculture, fisheries, offshore wind- and wave-power systems and other human interactions. The concept of assemblage is key here and merits sustained discussion, particularly with regard to one central aspect of the aquapelago that has largely escaped comment until now, its anthropocentricism. In accord with the very definition of *Human*ities as a field, the concept of the aquapelago – as elaborated to date – concerns human societies and interspecies and inter-material interactions initiated by them. Valid as this enterprise may be, it is important to identify that the imagination and discussion of integrated terrestrial and aquatic environments constituted by other species (beavers, for instance, who modify water flows through engineering; or coral or kelp, which create complex structures and environments for various species) is also a field that merits attention. Key to any such discussions is the concept of assemblage and of aquapelagos *being* assemblages. Within these, humans – while central – are only one of a series of actants who performatively constitute the aquapelago. In making this statement I draw on tenets of Actor-Network Theory, a body of thought that Bruno Latour acknowledges might better be entitled "actant-rhyzome ontology" (2005, 9). While the "rhyzome" (more usually spelt as "rhizome") referred here is the familiar multiple-noded entity proposed by Gilles Deleuze and Félix Guattari (1972/2004), the reference to "actant" merits more sustained discussion.

Put at its simplest, an actant is something that causes action. It can be animate (e.g., animals, plants, microbes, etc.), inanimate (minerals, gasses, etc.) or the manifestation of energy (storms, heatwaves, etc.) or systems (such as climate change). Actants become actants by performing an action, i.e., by impacting on other entities, some of which may already be performing an action and others that may be jolted into action by an actant. The aggregate of these actants and their activities constitute the assemblage central to the concept of the aquapelago. Humans who constitute aquapelagos through their engagements with terrestrial and aquatic spaces are (necessarily) engaged in interaction with what Jane Bennett describes as the "vibrant matter" of the environment, characterized by the "vitality" of various non-human things (2010, iii). Bennett's work is particularly pertinent for the concept since it elaborates the significance of these aforementioned aspects for understanding human society and – as the subtitle of her 2010 book specifies, "a political ecology of things" – through a "guiding question" that asks:

> How would political responses to public problems change were we to take seriously the vitality of (nonhuman) bodies? By "vitality" I mean the capacity of things – edibles, commodities, storms, metals – not only to impede or block the will and designs of humans but also to act as quasi agents or forces with trajectories, propensities, or tendencies of their own. My aspiration is to articulate a vibrant materiality that runs alongside and inside humans to see how analyses of political events might change if we gave the force of things more due.
>
> *(2010, iii)*

Bennett's central focus on political conceptualization and vision as the motive for her considerations parallels my own here. My concern to assert the concept of the aquapelago is not simply taxonomic but is rather a response to a number of questions of responsibility concerning how humans inhabit – and are causing and catalysing changes to – the planet, its oceans, its climate and its biomass. Aquapelagic assemblages are as impacted by the Anthropocene as any other. As sea levels rise, ocean warming, shifting currents and changes in biomass and biodiversity occur, those humans implicated in aquapelagic spaces are required to engage with a dynamic group of actants. While environmental changes have had to be negotiated from the earliest days of human existence, the period that most concerns me is the Anthropocene epoch, whose starting point is open to debate but whose full impact began to be apparent in the late 18th century. In this sense, my proposition of the aquapelago as a concept and focus is intended to facilitate comprehension of Anthropocene impacts on interrelated aquatic and land environments and of the impact on and responses of non-human actants. But the aquapelagic assemblages I propose as arising from the interaction of humans and other actants in particular locales are far from homogenous ones whose dynamics and results can be easily characterized; rather, I commend Bennett's characterization that:

> bodies enhance their power *in* or *as a heterogeneous assemblage*. What this suggests for the concept of agency is that the efficacy or effectivity to which that term has traditionally referred becomes distributed across an ontologically heterogeneous field, rather than being a capacity localised in a human body or in a collective produced (only) by human efforts.
>
> *(2010, 23 – emphases in original)*

Returning to this theme towards the conclusion of her study, Bennett offers what she describes as an "onto-story" (entitled "Natura Naturans" – a Latin term that can be translated as "nature doing what it does") (2010, 116–119). After contending that, "an affective, speaking human body is not radically different from the affective, signalling nonhumans with which it coexists, hosts, enjoys, serves, consumes, produces, and competes" (2010, 117); she goes on to assert that the environmental "field" these coexistent interactions occur within:

> is not a uniform or flat topography. It is just that its differentiations are too protean and diverse to coincide exclusively with the philosophical categories of life, matter, mental, environmental. The consistency of the field is more uneven than that: portions congeal into bodies, but not in a way that makes any one type the privileged site of agency. The source of effects is, rather, always an ontologically diverse assemblage of

energies and bodies, of simple and complex bodies, of the physical and the physiological.

In this onto-tale, everything is, in a sense, alive. This liveliness is not capped by an ultimate purpose or grasped and managed through a few simple and timeless (Kantian) categories.

(2010, 117)

What Bennett is evoking here is a generative assemblage that is "natural" in that it is diverse and multi-facetted, performed by multiple (and multiply interacting) actants. As will be apparent, the consideration of an aquapelago as such a lively assemblage in which humans interact with a range of other actants – sometimes imposing their will and/or causing unintentional impacts, sometimes blocked, diverted or defeated by interactions and reactions of other animates, inanimates or manifestations of energy – allows for insights into dynamic event phenomena within and between particular locations. Just as aquapelagos are performed entities, inhabitation of and mobility through them, and senses of them as home, require performances that engender the homeliness of such locales (Hayfield & Neilsen 2022).

In his response to my initial article on aquapelagos (Hayward 2012a), Ian Maxwell (2012) identified the concept of chorography as parallel to and supportive of the arguments around integrated island/marine spaces I advanced. Reflecting on critical emphases on "the primacy of space", he asserts that our beings have a "primal intercorporeality"; i.e., "rather than being set against the world we inhabit, we are given through and with it" and that we "live a radical continuity with our worlds" (2012, 23). This leads him to identify the late medieval concept of chorography, revived and re-inflected by writers such as Pearson (2007), as particularly pertinent. Chorography, in Maxwell's words, "renders (a) place in (its) chiasmatic idiosyncrasy, setting subjective and objective epistemologies into productive dialogue" (2012, 23). In accomplishing this, it also prioritizes a thorough recognition of and engagement with specificities. Chorography is, therefore, particularly congruent with aquapelagic analysis since each and every aquapelago is differently constituted and is temporally fluid. It is also congruent with Actor-Network theory. As Latour has characterized, while "Sociologists of the social seem to glide like angels, transporting power and connections almost immaterially" (2005, 25), those engaged in "actant-rhyzome ontology" have to engage with the specifics of places and the mesh of actants that perform them/in them at particular moments. Race, gender and class are also important elements here. Aquatic spaces do not dissolve social differences but, rather, inscribe and exploit them in terms of the types of humans employed to work in aquatic spaces in which work is arduous and often precarious (Guerin 2019, 2024; Lim, Ito & Matsuda 2012).

The temporality mentioned above is important. Aquapelagos are systems that come into being and wax and wane as climate patterns alter and as human socioeconomic organizations, treaties, laws, technologies and/or the resources and trade systems they rely on change and develop in these contexts. In this sense, aquapelagos are *performed* entities and their histories *as* performed entities can inform understandings of their present. This is particularly evident in the case of Newfoundland, whose European settlement arose as an adjunct to the maritime enterprise of the Grand Banks fishery, as detailed in the Introduction to Chapter 3 of this volume.

The following sections sketch the operation of two paradigmatic aquapelagic assemblages, The Torres Strait Islands and Haida Gwaii, in order to flesh out characterizations raised above. Both locales have been severely disrupted by Western colonialism but have succeeded in achieving some degree of restoration through engagement with the laws and institutions of settler powers that continue to manage them within colonial nation-state systems. The discussions that follow attempt to explore the broader implications of Pacific ways-of-seeing and being articulated in particular locations. While I discuss aspects of the Torres Strait and Haida Gwaii, in particular, I also draw on creative expressions of aquapelagality that I have encountered in various Pacific island contexts. A number of cultural texts and practices have brought home to me the deep senses of belonging that various island populations felt in marine environments and evoked these anew as I read, listened to and viewed them while drafting this chapter.

Exemplary Aquapelagos: The Torres Strait and Haida Gwaii

I first met Seaman Dan on Thursday Island (T.I.) with my colleague Karl Neuenfeldt in the offices of Radio 4MW. We had heard one of his impressive compositions – *T.I. Blues*, as recorded by the Mills Sisters – and we wanted to talk with him about his songwriting and performing career. He brought his guitar along and played a number of his songs for us[6] and told us about his experiences as a boat captain and pearl diver in the Torres Strait between the late 1940s and mid-1960s. He was modest, warm and happy to share his recollections of pearling days with us. One song that immediately grabbed my attention was *Forty Fathoms*,[7] whose lyrics referred to the Darnley Deep, an area off the coast of Erub Island where Seaman dived for pearl shell in his youth. The song sketches the precarity of diving at the titular depth (c80 metres) wearing the cumbersome suits typical of the period that were connected to surface vessels by long hoses. The lyrics also refer to Seaman seeing another besuited human working alongside him deep in the aquatic realm. The song exemplifies Ian Maxwell's discussion of how (various kinds of) performance "emplace place" in specific locales. Here it is a deep and potentially deadly pearl ground that attracted local workers on account of its offering one of the few local sources of paid labour in the region.[8] The pearl grounds were an archetypal aquapelagic space, generated at a specific time by particular livelihood activities, and they comprised a three-dimensional "stage" consisting of the seafloor, upon which the divers worked; the liquid space that they moved though, nervously, fearing decompression sickness ("the bends") if they rose too quickly; and the surface, where their mother ships awaited them. The immersive context and Seaman Dan's evocations of it from personal experience illuminated the distinct nature of aquapelagic assemblages.

The Torres Strait is a narrow stretch of island-studded water between Australia's Cape York Peninsula and the southern coast of Papua New Guinea (Fig. 1.1). While meeting all the common criteria for designation as an archipelago, the islands of the region are rarely referred to as such. As significantly, the common colloquial designation of the region is "The Straits"; a term that privileges the sea space between the two larger land masses as the defining element and echoes Hau'ofa by emphasizing its aquatic (rather than terrestrial) aspect. For the traditionally highly mobile marine people who exploited marine resources and traded across the seas between Australia and Papua New Guinea for centuries before Western colonization, this is of course apposite. Significantly, traditions of marine usage were a key element in the landmark Australian legal case that led to the rejection of the concept of *terra nullius*, formerly enshrined in Australian law as the pretext upon which Indigenous peoples were denied land rights due to the assumption that land was not "owned" in any way prior to European colonization of the continent. The landmark case was initiated by members of the Mer Island community, led by Eddie Koiki Mabo, in 1982 ("Mabo and Others v

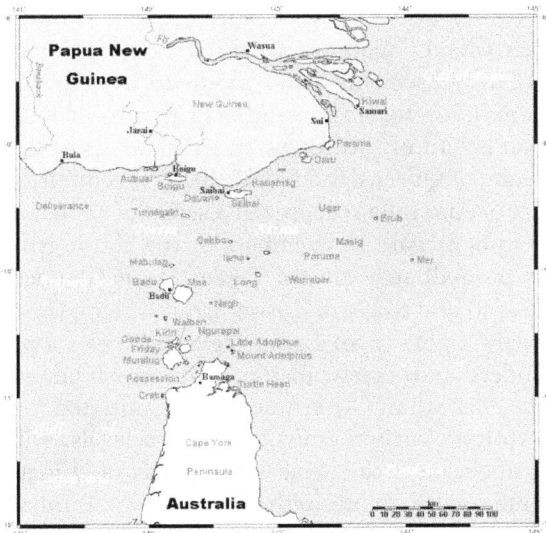

FIGURE 1.1   Map of Torres Strait and its islands (Wiki/GNU Public Licence).

Queensland n2"). After initially arguing for land rights on the island itself, Mabo extended his claim to two small reefs ten kilometers to the east of the island. Interviewed about the case for the film *Land Bilong Islanders* (1989, dir: Trevor Graham), Mabo identified the following cultural extension of Mer offshore:

> There is a stone fish trap that I'm claiming, and beyond that ... we have a lagoon that I call Las Kapar and beyond that again is our home reef called Op Nor. And then, of course, there is a stretch of sea which goes out to the Great Barrier Reef, and I claim that because it has special significance as far as our cultural myths and legends go.[7]

In all but use of the specific term, Mabo's statement points to an understanding of the Mer community as an aquapelagic society and of Mer itself as the terrestrial core of an aquapelagic assemblage.

The Australian Government opposed Mabo's claim, primarily on the grounds that traditional rights to marine areas were not congruent with their (Western) concepts of law and ownership, but, following a High Court hearing in 1992, the concept of *terra nullius* was formerly dropped from Australian law. Native title was subsequently recognized under the terms of the Australian Commonwealth's *Native Title Act* (1993), which led to a series of ongoing land claims. While the High Court stopped short of acknowledging traditional sea rights for Mer islanders, the Mabo case put the issue on the public agenda and began the momentum that led to the official recognition of Indigenous sea rights in 2001 when the Federal Court found in favor of the Croker Island community (located northeast of Darwin) and their claim for native title rights over two hundred square kilometers of sea floor to the east of their island ("Commonwealth v Yarmirr; Yarmirr v Northern Territory" 2001). A subsequent claim mounted by a collection of Torres Strait Island communities for marine rights in an area of 37,800 square kilometers ("Akiba on behalf of the Torres Strait Islanders of the Regional Seas Claim Group v State of Queensland No 2" 2013) was finally successful when the Federal Court of Australia recognized a series of rights that "were found to be possessed in aggregate by members of the claim group". While this represented only a small step toward Torres Strait Islander determination over marine access to and usage of Torres Strait waters, the official recognition of the islanders' rights to an aquapelagic space was a profound one. In recent years islanders' active stewardship of the region has extended to legal action against the Australian Government being initiated in 2021 over their minimal action to limit carbon emissions on sea levels, with the plaintiffs arguing that "if unchecked, the projected impacts of climate change in the Torres Strait would render islands in the Torres Strait uninhabitable, causing Torres Strait Islanders to become climate refugees and extinguishing Ailan

Kastom" [i.e. customary ways of island life] (Testa 2022 – also see Pabai 2022). Such concerns are shared across the Pacific.

In the period 1999–2001 I engaged in sustained research with the Pitcairn-descent population of Norfolk Island. Many of the songs written by islanders such as George Christian and Don Christian-Reynolds were striking on account of their expression of deep traditions and senses of community resilience in a remote island locale[9], but one song had a particular impact on my sense of terrestrial-maritime interrelations on the island – Kath King's *Teach me how fer lew*. Kath wrote the song in *Norfuk* creole[10] as an entry in a Norfolk Island song competition aimed to boost the production of locally themed compositions. Sung slowly to a pentatonic melody derived from the traditional English folk song *Scarborough Fair*, King's *Norfuk* lyrics were inspired by and describe her thoughts while standing on the cliffs above Black Bank, on the island's north-coast, and seeing a turtle swimming in the water below. The song relates an affective engagement between the singer and the turtle in which she specu-lates as to what turtles could teach humans about the history of the marine environment around the island, stretching back to the days when whaling was practiced there (between 1793 and 1907). As the verses progress, sketching different environmental scenarios, her reference to the turtle shifts, referring to it as "turtle ghos[t]" in verse three before characterizing the turtle as "the guardian of the sea" in verse 5. The song concludes with the line that gives the song its title, "teach ne how fer lew" (meaning 'teach me how to live'), with the previous verses lending that phrase the sense of how to live harmoniously with the natural environment. The sincerity of the lyrics, the direct, untutored quality of King's vocal delivery and its invocation of cultural traditions that had arrived on the remote island as a result of multiple voyages gave it both a time-*less* quality and a time*liness* as a comment on signs of ecological change. It was also striking for being sung with such evident love, conveying a deep emo-tional attachment to place and culture.

Having identified songs as especially moving and enlightening points of entry into aquapelagic sensibilities, I want to switch media and engage with an image that has preoccupied me since I first encountered it (but which is unfortunately not available for reproduction in this volume[11]). The image is the official crest of the Archipelago Management Board (AMB) of the Gwaii Haanas National Park-Reserve, National Marine Conservation Area Reserve and Haida Heritage site that occupies the southern portion of the Haida Gwaii archipelago, off the northwest coast of British Columbia (an area formerly referred to as the Queen Charlotte Islands). The image was painted by Haida artist Giitsxaa, from Skidegate, on Graham Island. The crest is rendered in a style typical of Haida and northwest coast Indigenous art, which uses delineating form-lines and ovaloid shapes with high degrees of symmetry and often ornate detail. Animals and mythological entities are rendered in stylized manners and there is a continuum of design styles between carved artifacts such as *gyáa'aang* (sometimes mischaracterized as "totem poles") and painted media. As well as being an accomplished example of Haida art in an aesthetic sense, what captured my interest was the temporal disjuncture between the image that the artwork represents and the actuality of biological components of the Gwaii Haanas NP-R that the Board manages. The image resonated with my memory of King's song *Teach me how fer lew*, and – in particular – her reference to the "ghos[t] turtle" that seemed to promise insights into ecological change and how it has affected various biological actors within the aquapelagic space. These issues are explicit in the crest's subjects, a reclining otter with a sea urchin on its belly. As with any image (compound or otherwise), there are various ways of interpreting it. One is to perceive the otter as floating on her back with the urchin on her belly as she readies to consume it. Sea urchins are a regular element of the diet of northwest Pacific otters and the image is not, therefore, that surprising.[12] But two aspects disturb this everyday scenario. Taking the infra-pictorial element first, the otter – while heavily stylized in the manner of Haida art – is recognizably a standard mammal. By contrast, the sea urchin's representation transforms the faceless, brainless and spiny, globular echinoderm into a being with a pseudo-mammalian appearance, with eyes, nose, mouth and facial expression. The faced sea urchin suggests sentience, rendering it less a spiny package with tasty internal organs ready to be consumed by a hungry otter than as a creature with agency that is actively engaged in its environment. The interaction is a frozen one whose imminent outcome may be suspected but is not yet known. Fitting, the (extra-pictorial) historical outcome of the interaction of the species was very different from the simple scenario suggested by the image. As discussed further below, while sea urchins are both present and, indeed, rampant around the islands, otters were eradicated from Haida Gwaii ecosystems over 100 years ago and have only recently started to return.[13] The image is, therefore, a complex one. When originally adopted, it evoked a past to haunt the present. But it can also be seen to have symbolically retained and maintained the otter over the species' interregnum in the archipelago, being invested with new meaning (and optimism) when sea otters began to return to local waters from the mid-2010s on and when the *X̱aayda Gwaay.yaay Ḵuugaay Gwii Sdiihltl'lx̱a* project was established to consider how to facilitate and respond to their return.[14]

Despite the erasure of the sea otters and the resultant ecological upheaval (described below), the crest also serves another purpose in representing the success of multi-partner initiatives in ensuring recognition of the aquapelagic nature of a region and establishing stewardship over it. As I go on to discuss, Gwaii Haanas's existence illustrates the manner in which the aquapelago can be used as a key referent in planning and administration of space drawing on Indigenous perceptions of the region, which have been summarized by the Council of the Haida Nation (2012, 1) with regard to Haida culture being "intertwined with all of creation in the land, sea, air and spirit worlds", and with "life in the ocean around us" as being "essential to our well-being" in that "it nourishes all of the communities of Haida Gwaii".

The present-day Haida Gwaii Islands (Fig. 1.2) were a product of rising sea levels following the end of the last Ice Age that resulted from the melting of the glaciers that covered the lowlands that now form the Hecate Strait between the islands and the continental coast of the Canadian province of British Columbia. Archaeology suggests a continuity of human settlement in the region dating back some 13,000 years and, since at least 2000 BCE, the Haida have comprised a group of well-established and powerful island-based clans that were reliant on coastal waterways to obtain various resources either through harvesting of marine materials or through trading with or plundering from other regional communities using large, durable seagoing canoes.[15] They also developed a sophisticated material culture most evident in large, elaborate, wooden structures that included complex and beautiful carved mortuary poles. The often-tempestuous waters around the

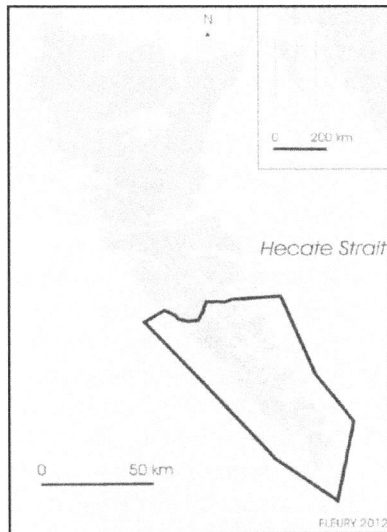

FIGURE 1.2   Map of Haida Gwaii aquapelago and Gwaii Haanas park (indicated by the bordered area to the south) and inset image of Haida Gwaii's position in regard to the coast of British Columbia. (Christian Fleury, 2012)

islands, particularly north toward the Prince of Wales Islands and across the Hecate Strait, required the development of stable, durable canoes and considerable navigation and crewing skills. While the waters adjacent to the islands may have been unpredictable in weather terms, the aquatic resources were plentiful, providing reliable all-year round foods such as fish (predominantly halibut and rockfish), shellfish (mussels, whelks, clams, etc.), aquatic mammals (seals and sea otters) and seabirds (mainly auks). In addition, salmon provided a valued seasonal resource and there is evidence of whales being consumed when these were washed ashore (Acheson 2005). The Haida appeared to have enjoyed regional security and supremacy from their island bases until the arrival of Westerners, initially in the form of explorers and traders and, subsequently, settlers who have modified the social fabric of the islands in various ways (Weiss 2019).

Available evidence suggests that at the time of initial contact with Western traders, the population of Haida Gwaii existed in a relatively stable equilibrium with their environment, inhabiting and enacting the aquapelago in an established and sustainable manner. This is, of course, not to suggest that the Haida had inhabited and exploited the area without environmental impact. Indeed, as Ian McKechnie and Rebecca J. Wigen have identified, the "direct use and long-term occupation of the region by coastal First Nations people … indicate[s] that humans have been participants in this ecosystem for at least the past 10,000 years and as such, likely directly and indirectly affected the distribution, growth, behaviour, and relative densities" of the marine resources they harvested (2011, 129).

The relative equilibrium described above did not last. Major disruption followed the arrival of Western traders in the late 1700s who were functionaries within a trans-Pacific trade network dedicated to acquiring sea otter pelts along the coast and islands of the Pacific Northwest and retailing them in China. Pelts were supplied by Indigenous communities in the Pacific Northwest and were traded for Western manufactured goods such as firearms, metal tools and manufactured cloth. Demand for the latter motivated the Haida to harvest pelts in the previously described intense and (as it proved) unsustainable manner, with otter numbers crashing to the point of local extinction around 1830 after fifty years intensive exploitation. The impact of the extinction of otters on fish stocks (through associated environmental factors referred to below) was largely obscured by another significant population decline, that of the human population of the region, which fell catastrophically due to the introduction and unhindered progress of hitherto unknown illnesses such as syphilis and smallpox. Between the mid-1830s and early 1880s the population declined by around 75% with smaller villages aggregating into single settlements and combining members of previously distinct communities. By the early 1900s these had further aggregated into two communities, Skidegate and Masset. Further upheavals followed in the form of a commercial salmon fishery, which severely depleted local stocks for the benefit of companies based on mainland British Columbia, and – most notoriously – logging.

Logging began in the islands in 1901 and expanded in the 1920s and 1930s with the harvesting of (lightweight but durable) sitka spruce. Logging continued through the 1950s and expanded rapidly in the 1960s–1980s, only levelling off after it had cleared many old-growth forests in the northern part of the islands and when it was increasingly opposed by anti-logging campaigns intent on protecting the pristine southerly areas of Moresby and adjacent islands. Local opposition to logging first became apparent in 1974, with the formation of the Islands Protection Committee in Queen Charlotte and via expressions of concern by the Skidegate (Haida) Band Council (Martineau 1999, 242). The Haida Nation established Gwaii Haanas as a Haida Heritage Site in 1985, the first step in the designations of the protected area. After over a decade of concerted direct action and negotiations with the Canadian and British Columbian Governments, the South Moresby Agreement was signed in 1988, leading to the establishment of a zone that became established as the Gwaii Haanas National Marine Conservation Area Reserve and Haida Heritage Site in June 2010. In 2018 this initiative was complemented by a regional Land-Sea-People Management Plan premised on the Haida concept of *Gina 'Waadlu̱xan Kil̲Guhl̲G̲a* (effectively, "considering everything because everything depends on everything else"). Its aquapelagic vision is clearly expressed in the statement that:

> Integrated management of the land, sea and people, considers the relationships between species and habitats, and accounts for short-term, long-term, and cumulative effects of human activities on the environment.
> *(Council of the Haida nation 2018)*

This is a striking vision since, as Gwaii Haanas's website identifies, in combination with the Haida Heritage site that comprises the southern part of Moresby Island and adjacent islands, the protected area "is currently the only place on Earth to be protected from mountain top to sea floor" and encompasses an ecologically rich and diverse marine terrain:

> Under the waters of the Hecate Strait, lie the contours of a former tundra like plain, with meandering rivers, lakes and beach terraces - a landscape drowned when sea levels rose after the last ice age. Off the west coast of Gwaii Haanas, the Queen Charlotte Shelf drops away abruptly to about 2,500 metres. This is an area of many transitions - between ocean abyss, continental slope, shallow shelf, and the dramatically upthrust landmass of the islands. Clean, nutrient rich water supports productive kelp forest communities and some of the most abundant, diverse and colourful intertidal communities found in temperate waters anywhere in the world.
> *(Council of the Haida nation 2018)*

I reproduce this statement since it is notable for combining a description of the various types and depths of ocean floor and of the biological communities that inhabit those regions and a strong sense of the (current) seafloor's retention of characteristics from the last glacial period (when it was a low-lying extension of the current island landscape). While largely neglected by Humanities-orientated island, coastal and maritime research, the relationship between seafloors (particularly heterogeneous ones), species diversity and human engagement with aquatic spaces is one that has only begun to be broached (e.g., Zeppilli, Pusceddu, Trincardi & Danovaro 2016) and merits further address,

Along with the aspects referred to above, the terms and clauses of the *Gwaii Haanas Agreement*[16] are noteworthy. The first key point is the recognition of Indigenous rights in the management of the area by the Archipelago Management Board (AMB). In addition to the expected stipulations of preserving natural resources, the agreement clearly enunciates a vision for the site that includes human presence in, and interaction with, spaces and animate presences in the region (Clauses 3.1, 3.2 and 3.3). It also asserts that:

> Matters to be addressed by the AMB will also include... identification of sites of special spiritual-cultural significance to the Haida within the Archipelago, including historic habitation and burial sites... and management of these sites on a case by case basis taking into account the requirements for protection of natural resources and cultural features, for Haida cultural activities, and traditional renewable resource harvesting activities... and for visitor understanding and enjoyment. (4.2c) [and] strategies to assist Haida individuals and organisations to take advantage of the full range of economic and employment opportunities associated with the planning, operation and management of the Archipelago.
>
> *(4.2h)*

Section 6 of the *Agreement* enshrines the following "cultural activities and sustainable, traditional renewable resource harvesting activities" as permissible within the site that include travel; gathering traditional foodstuffs, medicinal and ceremonial plants; logging for cultural purposes; hunting of mammals, freshwater and anadromous fish; "conducting, teaching or demonstrating ceremonies of traditional, spiritual or religious significance" and "seeking cultural and spiritual inspiration" together with the use of shelter and facilities essential to the pursuit of any of these activities.

From these viewpoints, it can be seen that the operating terms of Gwaii Haanas introduced and enshrined protection for an aquapelago constituted by the engagement of Haida with the locale that they inhabited prior to colonial incursion and disruption. The multiple provisions legally recognize the constituent elements that "organically" comprised a similar realm of human interaction with the natural environment in the pre-contact era. Like any such document, it necessarily compresses complexity and includes terms open for subsequent interpretation and dispute (an undefined reference to "sustainable" in Section 6 for example). The document, nevertheless, remains a notable aspirational charter that reflects what is – in all but name – an aquapelagic conceptualization of space and occupancy that attempts to reinstate a pre-Anthropocene regime in a modern spatio-administrative context.

Conclusion

The discussions of the Torres Strait and Haida Gwaii/Gwaii Haanas in the second part of this chapter serve to identify the importance of the aquapelagic vision outlined in the beginning of the chapter in two regards: (1) in attempts to protect aggregated island/marine environments from (further) Anthropocene degradation; and (2) to conceive of and legally encapsulate the complex relationship between humans and other animate and inanimate actants in integrated terrestrial and marine spaces. Crucially, it is the performance of aquapelagality that constitutes its entity. The aquapelago is a rich concept that is premised on human interaction with "vibrant matter" in particular locales. It is an "onto-tale" in which everything is interacting. It is not a product of a cartographic imagination, an image rendered flat. Indeed, it is the multiplicity of submarine depths, of regions of water and currents, of seafloor surfaces, of various forms of flora and fauna and their interactions with topologies of land and of aerial and weather systems as well as flows of materials between them that produces an aquapelago. Like any locale whose chorography can be understood historically, it is also a poetic space where traces and impacts of former interactions and former actants can be deployed to evoke what has been and gone and what may be in the future. These aspects are crucial to the human experiential, perceptual and representational practices that constitute aquapelagos. The aquapelago is a rich space, in all senses of that term. Its entangled terrestrial and aquatic flows and filaments weave a complex web through land and water. It is a complex holistic, multiple-actant environment in which terrestrial elements cannot be easily extricated from the assemblages that generate their social viability, character, imagination and potential futures. As Pugh and Chandler (2021, 14–15) identify, aquapelagic analysis is a "patchwork" enterprise that "develop[s] and transform[s] relational ontology" foregrounding "how entanglements of relation are never fixed". In these regards, the individual chapters in this volume variously represent and/or reflect on moments in the dynamic histories of aquapelagos from the points of view of ontologies that are themselves dynamic and subject to wide ranges of interpretation. The Anthropocene is crucial here. The intensity of climate change and weather events; related crises of human mobility, safety and security; manifestations of national self-interest and/or panic; and attempts to negotiate and counter these has accelerated event-time. The (relatively) short durée of the Anthropocene now witnesses accelerations of affect and events that the inhabitants and environments of aquapelagos and other socio-material entities have to deal with in increasingly shorter increments. Time is "out of joint" and the authors of the patchwork ontologies utilized in this volume have had to be nimble in order to represent and analyze places and events and suggest productive responses to Anthropocene change.

Acknowledgements

This chapter draws on, updates and extends material previously published in Hayward (2012a, 2012b) and responses to these debates in articles by subsequent authors in the journal *Shima*. Thanks to Jonathan Pugh, Miranda Post and Peter Moore for their feedback on the draft version of this chapter provided to them in early 2024.

Notes

1 As exemplified by organizations such as ISISA (the International Small Island Studies Association) and publications such as *Island Studies Journal.*
2 It was not until a decade later that a more credible approach to archipelagic studies/archipelagic ways of thinking began to emerge, largely premised on engagement with Hau'ofa and with Caribbean authors such as Édouard Glissant. See, particularly, Michelle Stephens and Yolanda Martinez-San Miguel's 2020 anthology *Contemporary Archipelagic Thinking* and the editors' insightful introduction (Stephens and Martinez-San Miguel (2020: 1–37).
3 Hayward (2012a, 2012b), Suwa (2012), Maxwell (2012), Dawson (2012), Baldacchino (2012), Nash (2013), Fleury (2013) and Cashman (2013).
4 The term *archipelago*, first recorded in English language use in the late 16th century, refers to an aggregation of islands. The name is commonly held to derive from two previous terms: *archi* from a Latin term meaning "chief" or "most prominent" and *pelagos*, from the Greek term for the sea. The expression appears to have entered anglophonic (and, hence, more global) usage from an Italian language term referring to a specific marine region, the Adriatic Sea; but its modern usage unambiguously refers to the land area of a group of islands *within* a sea. Our neologism thereby duplicates the etymology of the component *pelago* in order to emphasize the aquatic component of assemblages of terrestrial and aquatic elements.
5 See the dedicated online *Shima* anthology: https://shimajournal.org/anthologies /aquapelago.php#gsc.tab=0
6 Following this first encounter, Seaman Dan and Karl Neuenfeldt went on to pursue a fruitful collaboration that resulted in eight albums, two of which – *Perfect Pearl* (2004) and *Sailing Home* (2009) – won Australian Recording Industry Association Awards.
7 Included on his first album, *Follow the Sun* in 2000.
8 See Neuenfeldt (2002) for further discussion.

9 See Hayward (2006, pp 169–208) for a discussion of contemporary *Norfuk* language songs.
10 A mixture of English, Tahitian and other languages that originally developed on Pitcairn Island in the late 18th and early 19th centuries. See Hayward (2006) for a detailed history of Norfolk Island language and culture and Hayward (2006: 202–204) for a more detailed analysis of King's composition.
11 The crest is visible online at: https://gohaidagwaii.ca/about-haida-gwaii/
12 A range of photos of otters consuming sea urchins in this manner is available via a Google Images search for "otter sea urchin".
13 See N.A. Sloan and Lyle Dick (2012) for an insightful overview of the otters of Haida Gwaii and the pressure that led to their extinction.
14 See Parks Canada (2023) for a discussion of the otters' return and of the Gwaii Haanas and the Council of the Haida Nation *X̲aayda Gwaay.yaay K̲uugaay Gwii Sdiihltl'lx̲a* project initiated in 2020 in response.
15 See Moss (2008) for further discussion.
16 The *Gwaii Haanas Agreement* is a legally binding document under Schedule 2 of the Canada National Parks Act SOR 96/93.

References

Acheson, Steven. 2005. Gwaii Haanas settlement archaeology. In Daryl W. Fedje and Rolf Mathewes, eds. *Haida Gwaii: Human history and environment from the time of the loon to the time of the Iron People*. Vancouver: University of British Columbia Press: 303–336.

Ambon Statement. 2010. Concluding presentation to the International Conference on Small Islands and Coral Reefs (ISI-C), Ambon. *Shima* 4(2): 108–109.

Australian Government. 1993. *Native Title Act*. https://www8.austlii.edu.au/cgi-bin/viewdb/au/legis/cth/consol_act/nta1993147/

Baldacchino, G. 2012. Getting wet: A response to Hayward's concept of aquapelagos. *Shima* 6(1), 22–26.

Bell, Avril, Lythberg, Billie, Woods, Chris and Yukich, R. 2021. Enacting Settler Responsibilities towards Decolonisation. *Ethnicities* 22(5): 605–618.

Bennett, Jane. 2010. *Vibrant matter: a political ecology of things*. Durham: Duke University Press.

Cashman, David. 2013. Skimming the surface: Dislocated cruise liners and aquatic spaces. *Shima* 7(2), 1–12.

Council of the Haida Nation. 2012. *Ocean & way of life: Some things we know about Haida culture and the ocean and rivers of Haida Gwaii*. Council of the Haida Nation.

Council of the Haida Nation. 2018. Marine planning program: Gina 'Waadlux̲an KilG̲uhlG̲a - Gwaii Haanas land-sea-people plan. https://haidamarineplanning.com/initiatives/gwaii-haanas-land-sea-people-plan/

Dawson, Helen. 2012. Archaeology, aquapelagos and Island Studies. *Shima* 6(1), 17–21.

DeLaughrey, Elisabeth. 2001. "The litany of islands, the rosary of Archipelagoes": Caribbean and Pacific archipelagraphy. *Ariel: A Review of International English Literature 32*, 22–51.

Deleuze, Gilles and Guattari, Félix. 1972/2004. *Anti-Oedipus* (trans Robert Hurley, Mark Seem and Helen R. Lane). New York: Continuum.

Fleury, Christian. 2013. The island/sea/territory relationship: Towards a broader and three dimensional view of the aquapelagic assemblage. *Shima* 7(1), 1–13.

Garrison, R. 2019. Settler responsibility: Respatialising dissent in "America" beyond continental borders. *Shima 13*(2), 56-75

Graham, Trevor. 1989. *Land bilong islanders*. Documentary Film.

Guerin, Ayasha. 2019. Underground and at sea: Oysters and Black marine entanglements in New York's Zone-A. *Shima 13*(2), 30–55.

Guerin, Ayasha. 2025. We the submerged: (Non)Humans, race and aquapelagic relation - Notes from New York city. In Philip Hayward and May Joseph, eds. *Aquapelagos: Integrated terrestrial and marine assemblages*. Routledge.

Gwaii Haanas National Park Reserve and Haida Heritage Site Gwaii Haanas Agreement. 2010. www.pc.gc.ca/pn-np/bc/gwaiihaanas/plan/plan2/a.aspx#a4

Hau'ofa, Epeli. 1993. Our sea of islands. In Eric Waddell, Vijay Naidu and Epeli Hau'ofa, eds. *A New Oceania: Rediscovering our sea of islands*. Suva: University of the South Pacific, 2–16.

Hayfield, Erika Anne and Nielsen, Helene Pristed. 2022. Belonging in an aquapelago: Island mobilities and emotions. *Island Studies Journal 17*(2), 192–213.

Hayward, Philip. 2012a. Aquapelagos and aquapelagic assemblages: Towards an integrated study of island societies and marine environments. *Shima 6*(1), 1–11.

Hayward, Philip. 2012b. The constitution of assemblages and the aquapelagality of Haida Gwaii. *Shima 6*(2), 1–14.

Hayward, Philip. 2006. *Bounty Chords: Music, dance and cultural heritage on Norfolk and Pitcairn Islands*. Eastleigh: John Libbey & Co.

High Court of Australia. 1992. Mabo v Queensland (No 2) ("Mabo case") [1992] HCA 23; (1992) 175 CLR 1 (3 June 1992). http://www8.austlii.edu.au/cgi-bin/viewdoc/au/cases/cth/HCA/1992/23.html?stem=0&synonyms=0&query=title(mabo%20%20near%20%20queensland)

High Court of Australia. 2002. Commonwealth v Yarmirr [2001] HCA 56; 184 AJR 113; 208 CLR 1; 75 ALJR 1582 (11 October 2001). http://www8.austlii.edu.au/cgi-bin/viewdoc/au/cases/cth/HCA/2001/56.html?context=1;query=yarmirr;mask_path=au/cases/cth/HCA

High Court of Australia. 2013. Akiba on behalf of the Torres Strait Regional Sea Claim Group v Queensland [2010] FCA 321. http://www8.austlii.edu.au/cgi-bin/viewdoc/au/cases/cth/HCA/2013/33.html

Latour, Bruno. 2005. *Reassembling the social: An introduction to actor-network theory*. Oxford: Oxford University Press.

Lim, Cristina P., Ito, Yasuhiro and Matsuda, Yoshiaki. 2012. Braving the Sea: The Amasan (Women Divers) of the Yahataura Fishing Community, Iki Island, Nagasaki Prefecture, Japan. *Asian Fisheries Science 25S*, 29–45.

Martineau, Joel. 1999. Otter skins, clearcuts, ecotourists: Re-resourcing Haida Gwaii', in Marc L. Miller, Jan Auyong and Nina P. Hadley (eds.) *Proceedings of the 1999 International Symposium on Coastal and Marine Tourism: Balancing Tourism and Conservation*: 237– 49.

Maxwell, Ian. 2012. Seas as places: Towards a maritime chorography. *Shima 6*(1), 27–29.

McKechnie, Ian and Wigen, Rebecca J. 2011. Towards a historical ecology of pinniped and sea otter hunting traditions on the coast of southern British Columbia. In Todd J. Braje and C. Rick Torben, eds. *Human impacts on seals, sea lions and sea otters: Integrating archaeology and ecology in the Northeast Pacific*. Berkeley: University of California Press, 129–166.

Moss, Madonna L. 2008. Islands coming out of concealment: Travelling to Haida Gwaii on the Northwest Coast of North America. *The Journal of Island and Coastal Archaeology 3*(1), 35–53.

Nash, Joshua. 2013. Naming the aquapelago: Reconsidering Norfolk Island fishing ground names. *Shima 6*(2), 118–131.

Neuenfeldt, Karl. 2002. Torres strait maritime songs of longing and belonging. *Australian Studies 26*, 111–116.

Pabai, Pabai. 2022. Rights-holders from Torres Strait sue Commonwealth over climate change. *Native Title Newsletter 1*, 2–4.

Parks Canada. 2023. X̱aayda Gwaay.yaay K̲uugaay Gwii Sdiihltl'lx̲a: The sea otters return to Haida Gwaii. https://parks.canada.ca/pn-np/bc/gwaiihaanas/nature/conservation/restauration-restoration/kuu

Pearson, Mike. 2007. *'In comes I': Performance, memory and landscape*. Exeter: University of Exeter Press.

Popova, Ekaterina, Aksenov, Yevgeny, Amoudry, Laurent, et al. 2023. Socio-oceanography: An opportunity to integrate marine social and natural sciences. *Frontiers in Marine Science 10*. https://www.frontiersin.org/articles/10.3389/fmars.2023.1209356/full

Pugh, Jonathan and Chandler, David. 2021. *Anthropocene islands: Entangled worlds*. London: University of Westminster Press.

Sloan, N.A. and Lyle, Dick. 2012. *Sea otters of Haida Gwaii: Icons in human-ocean relations*. Skidegate: Archipelago management Board and Haida Gwaii Museum.

Stephens, Michelle and Martinez-San Miguel, Yolanda, eds. 2020. *Contemporary archipelagic thinking: Towards new comparative methodologies and disciplinary formations*. Lanham: Rowman and Littlefield, 1–44.

Stephens, M. and Martinez-San Miguel, Y. 2020. "Isolated above, but connected below": Toward new, global, archiplegic linkages. In M. Stephens and Y. Martinez-San Miguel, eds. *Contemporary archipelagic thinking: Towards new comparative methodologies and disciplinary formations*. Lanham: Rowman and Littlefield, 1–44.

Suwa, Jun'ichiro. 2012. Shima and aquapelagic assemblages: A commentary from Japan. *Shima 6*(1), 12–18.

Testa, Christopher. 2022. Judge wants timely climate trial as Torres Strait Islanders watch 'march of the sea'. *ABC Far North*, July 2022. https://www.abc.net.au/news/2022-07-23/torres-strait-islander-climate-trial-set-for-federal-court/101261774#:~:text=Pabai%20Pabai%20and%20Paul%20Kabai%2C%20native%20title%20holders%20from%20the,effects%20of%20greenhouse%20gas%20emissions.

Weiss, Joseph. 2019. *Shaping the future on Haida Gwaii: Life beyond settler colonialism*. Vancouver: University of British Columbia Press.

Zeppilli, Daniela, Pusceddu, Antonio, Trincardi, Fabio and Danovaro, Roberto. 2016. Seafloor heterogeneity influences the biodiversity–ecosystem functioning relationships in the deep sea. *Scientific Reports 6*, 26352.

2

SHIMA, SHIMAGUNI AND AQUAPELAGIC ASSEMBLAGES

Jun'ichiro Suwa

Introduction

This chapter explores how the idea of the aquapelagic assemblage is manifest in Japanese cultural landscapes. As I have written elsewhere (Suwa 2005), the Japanese concept of *shima* ably reflects the idea of the aquapelago in that *shima* is an assemblage of the geographic characteristics of islands and their waters and is inseparable from human activity. In the Japanese and Okinawan lexicon, *shima* not only means "island", *per se*, but also communities and neighborhoods on land. Therefore, the idea of *shima* suggests that the scope of the aquapelago must take account in human activity.[1] To borrow a word from Anna Lowenhaupt Tsing's 2021 monograph about *matsutake*, this activity is a "disturbance". Since the aromatic *matsutake* mushroom grows on oligotrophic pine forests with poor undergrowth, during the time when dead pine leaves and twigs were used for firewood and the trees were cut for building materials, it was more commonly available that it is today. The aquapelago is like the *matsutake* in that constant "disturbance" by means of cultural activity is essential for its sustainability. The Nan Madol ruins of Pohnpei (Micronesia) also illustrate that human "disturbance" of the stone structures and megaliths, which were built on artificial ground above the sea level by prehistoric people, thereby creating a new micro-ecosystem. Neolithic human activity on Nan Madol must have taken place in the artificial city with certain sophistication and, even after its abandonment, tourism and academic interests maintain the site as an aquapelago.

In reality, the aquapelago is complex in that *shima* can operate as a political space as well. This is further entangled when the idea of nation-state

DOI: 10.4324/9781003569534-2

overlaps with or takes it over. In Japanese, the word *shimaguni* signifies *shima* as a politically coded space. *Shimaguni* – "island nation" or "island country" – is a product of a "mythical reality" of the aquapelago that symbolizes the livelihood of *shima* as a homeland or a nation. The entanglement takes place when *shimaguni* and *shima* operate on the same geographic entity simultaneously, as exemplified by the Japanese archipelago. The fundamental difference between *shima* and *shimaguni* is not so much a matter of dimension and scale but rather that of the *modus operandi* of symbols or representations, as well as substances that materialise and effectuate such symbols. Whereas *shima* directly involves the commons, *shimaguni* expresses the identity of the homeland and, therefore, cannot be devoid of external political significance and discourse. *Shimaguni* constitutes an aquapelago in terms of the political interaction of apparatus or space produced by a regime of symbols. *Shimaguni* concerns the identity with which the aquapelago is practiced, imagined and historically enforced. Hence, this chapter will start by discussing the concept of *shimaguni* in detail and then come back to the notion of *shima* as aquapelagic assemblage.

Shimaguni

It is not certain in which period the word *shimaguni* started to be used. However, the composite word signifies the Japanese nation as a counterpart against its continental neighbors, most notably the classical Sinic civilization of East Asia. The term *shimaguni* is a composite of *shima* and *kuni*[2] [also renderable as *guni*] ("country", "nation" or "homeland"), rendered in kanji by the juxtaposition of the two words alongside each other (rather than by a single pictograph) thus:

$$島国$$

In contemporary Japan, while the words *shima* and *shimaguni* coexist with different nuances, they can also overlap with each other. This means that the 'imagined community' (Anderson 1983) in modern Japan is deeply embedded in the figure of the aquapelago (Suwa, 2012). However, the identification of islands as a nation also existed prior to the modern era. The best-known example would be *Kojiki*, the oldest surviving compendium of myths and imperial chronicles from the 8th century, which narrates the Japanese archipelago as the body of the nation (Suwa 2021). The volume starts with the origin myth of *shimaumi* ("island-bearing"), depicting that the islands, among various other natural entities, were delivered from the womb of the creatrix Izanami. The goddess and her husband Izanagi created the first island of Onokoro from the murky primordial chaos into which they descended. The myths in *Kojiki* appear to have been forgotten for hundreds of years until a number of scholars attempted to interpret the text in the latter half of 18th century, a time when the *literati*, however slowly, became concerned about the future of Japan in the face of Western colonialist expansion in Asia.

The idea of Japan as an island nation, as widely known today, however, is a relatively recent one. The ancient Japanese state was modelled on the centralised feudal systems of mainland Asian nations such as China and Korea. Inspired by these continental entities, Japan imagined itself in similar terms and overlooked its archipelagic composition. As a result, the names of Japan's four main islands, being the domain of political centres, do not contain the suffixes –shima (-*jima*) or –*tō* to indicate their islandness. Such designations were reserved for the margins. In the north of the archipelago, Hokkaido was formally called Watarishima ("crossed island") or Ezogashima ("The island of the Ezo"[3]); and in the south, the Amami Islands were called Michinoshima ("trail islands"). This sense of the outer islands' peripherality led them to be used as places of exile for internal dissidents, resulting in the terms *shima-okuri* or *shima-nagashi* (literally "throwing away to an island") being used to refer to political exile. Traditional Japanese art also reflected the idea of islands as isolated and detached from the cultural and political centre. A Noh theatre masterpiece entitled *Shunkan*, for instance (attributed to Zeami Motokyo), was inspired by an episode in the *Heike Monogatari* ("Tale of the Heike clan") about the exile of the monk Shunkan to Kikaijima.[4] Shunkan's story inspired many subsequent versions and was also adapted into a *jōruri* (puppet play) as well as a *kabuki* text. As these examples illustrate, small islands played their part in the territorial imagination of pre-modern state centres such as Kyoto and Edo.

After Japan's contact with the Jesuits in the 17th century, old maps introduced from the Mongolian Empire became updated and widely available and people discovered the location and proportion of their island nation. As a result, Japan's internal image as a political centre was offset by its representation as an insular periphery.[5] The well-known aphorism attributed to 18th-century intellectual and military scholar Hayashi Shihei that 'between Nihonbashi Bridge [in Tokyo] and Europe, there is only one waterway' exemplified debate around Japan's isolationist policy in the period and the country's surrounding waters were subsequently recognised as an important aspect of national security. The consciousness of the Japanese state as a *shimaguni* (island nation) was one that implicitly recognised the negative connotations that might accrue to it in terms of its potential to be regarded (by the West, for instance) as insular, reclusive, inward, backwards, primitive, narrow-minded, over-crowded, resource-scarce and so on. Continental Asia, on the other hand, was often abbreviated as *tairiku* ("the continent") and was regarded as a region ripe with opportunity. In the early 20th century, the media romanticised lonesome *tairiku-ronin*[5] (itinerants who sought their fortunes in China), as Japan's national interest turned to colonial expansion. The geopolitical consciousness of feudal-era Japan was reversed in imperialist Japan as its focus turned outwards. The *naichi*, the "inner (is)lands" of Japan became the central point, and continental Asia became its *gaichi* ("outer lands").[6]

Since World War II, the term and concept of *ritō* ("remote islands") has served as a key trope in the imagination of Japan's islands, especially as depopulation became a serious issue in the 1950s. A number of programmes were planned to "de-*ritō*-nize" islands and lawmakers and policymakers and the national public continued to conceive the islands within established mainland-satellite or centre-periphery dichotomies. One manifestation of this was the use of central Japanese funds to connect islands via bridges, thereby modifying their status as islands. However, some intellectuals and activists attempted to approach the *ritō* issue from a different perspective. Miyamoto Tsuneichi, for example, who hailed from Suō Ōshima Island himself, visited dozens of remote islands to record their folklore and called for public and political attention to promote their cultural survival (Miyamoto 1960).[7] Novelist Shimao Toshio, a long-time resident of Amami, also originated the term "Yaponesia"[8] in an attempt to decentralise the power relationship between the (notional) centre and periphery and to stimulate related post-colonial critiques of the national entity (Shimao 1977).

In contemporary Japan, the national self-image as a *shimaguni* ("island country") is premised on a mixture of political and indigenous images of territory. In fact, *tō*, a loan word from classical Chinese, only appears as a prefix or suffix, often deployed in order to compose academic jargon. The indigenous idea of *shima*, which would sound more natural in everyday conversation, connotes the island as a lived world; it signifies not simply a piece of land, but a space generated by livelihood or cultural conduct. A community or neighbourhood can also be called a *shima*, and not as a metaphor. For instance, to call an area of a city a *shima*, which is possible, is to emphasise that the place is one's hometown, neighbourhood or territory. In this regard, *shima* signifies livelihood or a sphere of influence. It is fractal, since the centre of livelihood activity produces concentric circles from a household to the globe. *Shima* is a spatio-temporal concept and a work of imagination where landmarks generate a sense of reality (Suwa 2005). A *shima*-esque archipelago forms concentric circles with others, and the centre can be collective or individual, or human or non-human insofar as the concentric circles may be imagined to weave cultural, ecological, climatic and geophysical interactions. The concentric circles of *shima* in this regard form an assemblage or rhizome (Deleuze & Guattari 1987). In an assemblage, there is no absolute centre that dominates the rest; rather, everything participates with and/or comprises the other. There is no distinction between parts and the whole, or any centre-periphery dichotomy. In such assemblages, concentric circles are effectively decentralised because there will be multiple centres with all sorts of behaviours, movements, perspectives and vanishing points. One's centre is simultaneously someone else's periphery, and one's centre is also someone else's within a different frame of meaning and significance.

Shima and Aquapelagic Assemblage

The concepts of the aquapelago and aquapelagic assemblages that Philip Hayward (2012) proposed are interesting for introducing the idea of assemblage to the consideration of island regions. If the assemblages are regarded in terms of the inseparability of water and land, such areas can be visualised as rhizomes. In this sense, an aquapelago is also a fractal because land and water are merging to constitute a whole. In aquapelagos, islands are not groups of isolates but rather assemblages concentrated by the waters. The relationship between islands, islands and continental mainlands, land and water, etc., are inseparable, as they are linked with each other in ways that can be grasped only via an invertible or intersubjective concept that "sea is land and land is sea". Although this might sound somewhat oxymoronic, the distinction between the land and the sea is merely a product of imagination. There are types of livelihood that do not see them separately. To the Bajau of the Sulu Sea, the Moken of the Andaman Sea and the residents of *ebune* houseboats who pursue a traditional lifestyle in some parts of coastal Japan – to name but a few examples – aquapelagic assemblages are an everyday reality where the distinction between land and sea becomes nonsensical. Similarly, among communities around Madang (Papua New Guinea), commuting between islands and the mainland by canoe has become so common that it can be compared with driving a vehicle (indeed, in the local Bel languages the term *wag* initially meant a small canoe but today refers to any form of transport such as a car or a bus).[9]

With regard to the idea of aquapelagic assemblages, the concept of "assemblage", as discussed by Gilles Deleuze and Félix Guattari, merits further attention. The assemblage of (human) livelihood provides a mediating process between land and marine environments and it constructs cultural landscapes with diversity as well as complexity. People hunt, gather or grow products from the sea as well as from the land, and when their "commons" are open, wildernesses and grasslands can be understood as a "sea". In discussing aquapelagic assemblages, therefore, the process by which cultural landscapes are generated turns out to be a central issue; they are assemblages and the distinction between sea and land evaporates in the reality of everyday life. Aquapelagic assemblages therefore cannot be discussed as a single mode of production.

One of the possible ways of grasping the nature of aquapelagic assemblages might be to focus on locations where fractal concentric circles initiate activities. This point might be regarded as a "sanctuary", using an expanded definition of that term that includes its complementary characterisations of a holy/spiritual centre, a place of refuge and safety and a reserve where flora and fauna are protected. This concept is similar to *kami no commons*, the notion of "commons" conceived by Akimichi Tomoya (2004) that derives from the sanctuary offered by Shinto shrines. These shrines are open commons in the sense that almost anyone can enter into their domain but at the same time they can also be regarded as restricted since spiritual beings rule over and imbue the space and determine the environment. The closure involved is conceptual/spiritual and is manifest in the conservation of trees, mountains, rocks, water sources (and, sometimes, even an island) intrinsic to the shrine. Conceived in this way, sanctuary comprises a local commons where concentric circles interact with degrees of inaccessibility; as in the way that religious sites and monuments form a centre in a neighbourhood. In aquapelagic assemblages, various levels, from local ecosystems to belief systems, are relevant and ownership adds complexity in forming assemblages that range from collective to private property.

Aquapelagic assemblages can be conceived as sanctuaries with regard to either real or imagined spaces; in fact, whether they are imagined or real is relatively insignificant. In some Shinto rituals, a sacred float called *mikoshi* is carried from shrines to the sea in order to rejuvenate the spirit. Here, the god on the float is imagined but the procession of the float is real, and the movement of the float to the coast reflects the imagined sanctuary of aquapelagic assemblage through its ability to maintain the environment. In Okinawan mythology, *Nirai Kanai*, the domain of gods, from which all life originates, is located over the western horizon and community religious sites such as *utaki* are plexuses of divine and human activities. The ideas of sanctuary explicated here can also be extended beyond traditional or religious beliefs and include contemporary concepts such as the conservation of the natural environment or heritage and/or sustainable development programs.[10] Key to the imagination of sanctuary is a shared idea that constructs cultural reality and thereby becomes crucial to any such project. Sanctuary therefore is a type of spatio-temporal space where things take place, operate and interact within a particular framework. In aquapelagic assemblages, landscape and personhood merge and interact. The making of *shima* can involve ritual and performance, as in the case of Benten worship in Shimokita Peninsula (Suwa 2017). This barren rock, now attached to a wharf is named Bentenjima (the "island of Benten"), to enshrine the goddess of fishery and wealth. During an annual festival, the villagers look at the mainland from the rocky sanctuary where their fellows conduct rituals and perform folk dances. This ritualistic character of the aquapelago, which shifts experiences, extends to artworks such as those open-air exhibitions on Teshima. Here, the wind blowing down the bushy mountain becomes a work of art in Christian Boltanski's *Forest of Murmurs* and the sound and soft wind from flowing springwater in a Shinto shrine resonates with Noe Aoki's *Particles in the Air*. Furthermore, these open-air artworks, which are located on roadsides, generate connections with the route of Buddhist pilgrimage through which the local way of life is embedded (Suwa 2020).

The constitution of aquapelagic assemblages involves the appropriation and sharing of cultural land/seascapes, therefore occupancy of space is a primary consideration. Aquapelagic assemblages – and/or the sanctuaries they comprise – are not necessarily self-contained, self-sufficient and/or self-sustainable, they are specific products of ongoing processes in actual locations; and the resource uses, finances, identities and alliances involved in constituting them are fluid, transient and sometimes elusive. Aquapelagos are sanctuaries manifested in land/seascapes, and their domains are occupied with sacred, untouchable, memorised and/or identified elements that are shared and as common knowledge, skill, consciousness, desire and ideology, whatever the direction they take. Understanding *shima* as aquapelagic assemblages depends on the sharing of sanctuary. What appears to be collective or communal ownership is in fact the reverse; *shima* are not the subject of ownership but the product of sanctuary in practice.

Conclusion: Between *Shima* and *Shimaguni*

The 19th-century interpretation of *Kojiki* developed contemporary ideas of the Japanese archipelago by means of its mythical episodes. The fact that myths play such an important role indicates that the complex relationship between *shima* and *shimaguni* provides an epistemological link to aquapelagic assemblages. Myth, legend and fantasy should not be treated as a product but as a process of perception, orality, performance or any other ways of grasping worlds. What we need to see here is how exactly such an "elusive" mindset interacts as a mode or style of *pensée sauvage* ("untamed thought") as represented in Blake's famous coloured etching "The Ancient of Days".[11] This chapter has discussed the aquapelago as an assemblage of *shima* and *shimaguni* and has attempted to explain that the former is the generator of the commons and sustainability while the latter operates to demarcate, enclose, fence, create barriers and exclusive articulations. This indicates that *shimaguni* does not simply represent the adoption of the idea of the nation compounded with *shima*. *Shima* is not a primordial version of *shimaguni*, and *shimaguni is* not a codified canonical abstraction of *shima* and did not evolve from *shima*. The core question is why and how these two different spaces can exist in the same framework. Further, at least in the example given in this chapter, *shimaguni* does not rely on the essence of *shima*, in a larger sense, with regard to the generation of the reality of *shima*, or the aquapelago. Rather, the two are the substance of "*shima* reality" since they always appear as something which is felt, told or practiced. By way of illustration, Okinotorishima, an atoll located some 17,000km south of Tokyo has two tiny rocks named Higashikojima and Kitakojima that are protected from erosion by concrete casings in a lagoon,[12] which evoke the *Kojiki* creation myth. The gods Izanagi and Izanami are believed to have created Onokoro Island to initiate lives on the earth. Whereas the exact location of Onokoro Island is unknown, the myth substantiates that at least one *shima* somewhere in central Japan is/was the place of *shimaguni*'s origin. The act of generating and remembering Onokoro and cementing the rocks of Okinotorishima substantiate an assemblage as an aquapelago, where the land and waters are identical with life.

Acknowledgement

I especially thank Phillip Hayward for his deep encouragement to me to explore this fascinating field of study.

Notes

1 Folklorist Yanagita Kunio hypothesised that the word *shima* originally meant a settlement, a compound, a group of homesteads or a village. *Shima* subsequently became a synonym for 'island' as settlements were often developed on uninhabited small islands (See Yanagita 1961).
2 It appears that *-ni* means soil or earth in ancient Japanese, as in *aoni* ("mud" or "verdigris") or *haniwa* (i.e. *haniwa* terracotta).
3 Perhaps from *enchu* or *enchiu* ("people") in Ainu. *Suwa Daimyōjin Ekotoba*, an old chronicle of the Suwa Taisha shrine, which was completed in 1356, refers to *Ezo-ga-chishima* (A thousand islands of Ezo). This means that Hokkaido was initially recognised as an aquapelago which was home to non-Japanese-speaking groups.
4 Ironically, *Kikaijima* had a *gusuku* state that flourished in the 9th century. Archaeological research suggests that the ancient state prospered as a hub of East Asian trading network.
5 This took place in parallel with imagining the aquapelago as a chain of islands overseas. Today, the word *shima* also means "stripe" because striped cloths were imported from Southeast Asia and called *shima-watari* ("came across the islands"). *Bengara-jima* and *Santome-jima* are named after Bengal and São Tomé, today's Chennai. These striped designs were favoured by kabuki actors and became popular.
6 Philosopher and critic Tsuchida Kyōson published a book entitled *Japan's Future as shima-kokka* (1924). Despite the fact that Japan had already annexed Korea (1010) and operated colonial enclaves in China (1905), Tsuchida still persists in characterising the nation as *shima-kokka* (island-state).
7 Miyamoto was the key figure in the legislation of the *Ritō Shinkō Hō* (Island Development Act) of 1953 Similarly, the *Hantō Shinkō Hō* (Peninsula Development Act) followed in 1980. This indicates that peninsulas and islands share similar characteristics in Japanese politics.
8 *Shima*'s neologism is sometimes spelled as "Japonesia" for it is a composite of Japonia/Iaponia (Latin/Greek) and *nesia* (Classical Greek). However, the romaji spelling of "Yaponesia" preserves the original sound and should be favoured.

9 The image of the canoe also refers to the dead as they appear in a number of
guitar band song lyrics (Suwa 2001).
10 See Hayward (2024) elsewhere in this volume for discussion of the submarine
sanctuary of the Flower Garden Banks.
11 Online at: https://www.britishmuseum.org/collection/object/P_1859-0625-72
12 See Tsuruta (2023).

References

Akimichi, T. 2004. *Commons no Jinruigaku*. Tokyo: Jinbun Shoin.
Anderson Benedict. 1983. *Imagined communities: Reflections on the origin and spread of nationalism*. New York: Verso.
Deleuze, Gilles and Guattari, Felix. 1987. *A thousand Plateaus: Capitalism and schizophrenia*. Minneapolis: University of Minnesota Press.
Hayward, Philip. 2012. Aquapelagos and aquapelagic assemblages. *Shima* 6(1), 1–10.
Hayward, Philip. 2024. The Flower Garden Banks and the parameters of aquapelagic sanctuary. In Philip Hayward and May Joseph (eds.) *Aquapelagos: Integrated Terrestrial and Marine Assemblages*. New York: Routledge, xx–xx.
Kojiki. 1968. Translated by Donald L. Philippi. Tokyo Press. https://archive.org/details/kojikitranslated00phil/page/n7/mode/2up
Miyamoto, Tsuneichi. 1960. *Nihon no Ritō*. Tokyo: Mirai Sha.
Shimao, Toshio. 1977. *Yaponesia-kō*. Tokyo: Ashi Shobō.
Suwa, Jun'ichiro. 2001. Representing sorrow in stringband laments in the Madang area, Papua New Guinea. *People and Culture in Oceania 17*, 47–66.
Suwa, Jun'ichiro. 2005. The space of shima. *Shima* 1(1), 6–15.
Suwa, Jun'ichiro. 2012. Shima and aquapelagic assemblages. *Shima* 6(1) 12–16.
Suwa, Jun'ichiro. 2017. Becoming island: The aquapelagic assemblage of Bentensai Festival in Sai, Northern Japan. *Shima 10(2)*, 5–19.
Suwa, Jun'ichiro. 2020. Artwork, assemblage and interactivity on Teshima. *Shima 14(2)*, 231–249.
Suwa, Jun'ichiro. 2021. Shimaumi: Aquapelagic imagery and poetics of 'island-laying' in Kojiki. *Coolabah 31*, 67–79.
Tsing, Anna Lowenhaupt. 2021. *The Mushroom at the end of the world: On the possibility of life in capitalist ruins*. Princeton: Princeton University Press.
Tsuchida, K. 1924. *Shimakokka toshiteno Nihon no mirai*. Tokyo: Naigai Shuppan.
Tsuruta, Jun. 2023. China's navy, the West Pacific and the role of Okinotorishima. *The Diplomat*. https://thediplomat.com/2023/01/chinas-navy-the-west-pacific-and-the-role-of-okinotorishima/
Yanagita, Kunio. 1961. *Kaijō no michi*. Tokyo: Chikuma Shobō.

3

MAKING AQUAPELAGIC PLACE IN JERSEY

The Écréhous and Minquiers Reefs

Christian Fleury and Henry Johnson

Introduction

This chapter outlines the making of aquapelagic place in the Bailiwick of Jersey (hereafter Jersey), a British Crown Dependency in the Gulf of Saint-Malo to the northwest of France, with specific reference to the Écréhous and Minquiers reefs (Fig. 3.1).[1] Jersey is not a constituent part of the United Kingdom. It possesses its own autonomous government and exercises judicial independence. However, it maintains strong constitutional ties to the United Kingdom by virtue of its relationship with the British Crown (States of Jersey 2024). The discussion is framed around the concept of "aquapelago", defined as "an assemblage of the marine and land spaces of a group of islands and their adjacent waters" (Hayward 2012a, 5), and, in particular, as a form of "vibrant matter" (2012a, 12) involving multiple interactions across changing dimensions of existence (Hayward 2012b), In this chapter, we show how aquapelagic place is constructed across horizontal and vertical dimensions, based around not only the political contestation of international borders and sub-national jurisdiction, but also within the layered spheres of traditional fishing rights, environmentalism and tourism. We are particularly intrigued by the concept of livelihood activities playing a pivotal role in the establishment of aquapelagos, expounded by Jun'ichiro Suwa in the following terms:

> In an assemblage, there is no absolute centre that dominates the rest rather, everything participates with and/or comprises the other. There is no distinction between parts and the whole, or any centre-periphery dichotomy.
>
> *(2012, 14)*

DOI: 10.4324/9781003569534-3

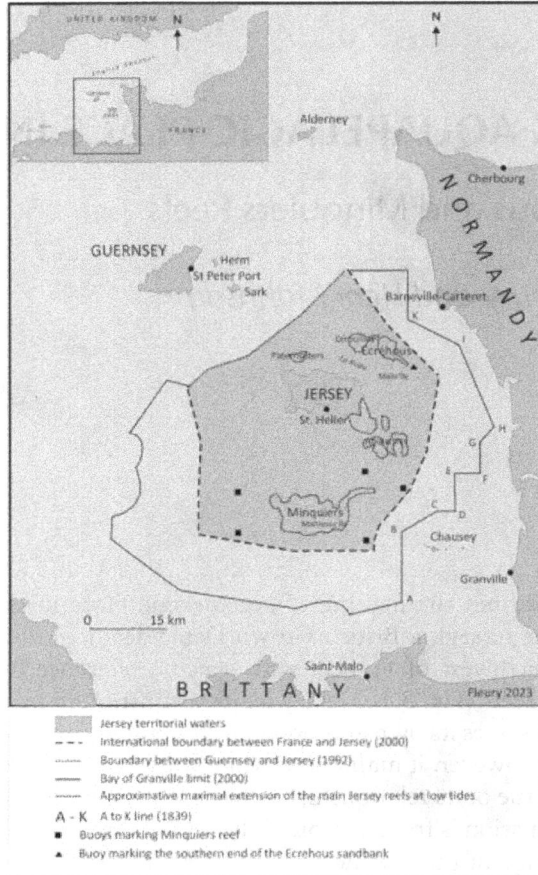

FIGURE 3.1 Jersey and its territorial waters. (Christian Fleury)

Jersey has several distinct offshore reefs located within its shallow seas (about 10–30m deep). The two main reefs that have a history of low-level anthropogenic influence are the Écréhous, which lie to the northeast of the island and include the smaller reefs to the west, Les Dirouilles and Les Pierres de Lecq (Paternosters), and the Minquiers, which are situated to the south of Jersey. Another reef system, Les Anquettes, lies to the southeast of the island, although it is less exposed than the other reefs.

The Écréhous and Minquiers have been chosen for discussion because of their history of settlement (short term and temporary) and having the largest islets amongst Jersey's reefs. Specifically, we place the reefs within a dynamic and transitional context, where various social domains intersect, creating a terrestrial and aquatic space that bridges a British archipelago with the French mainland, thus giving rise to a sense of place that emerges from both land and sea. Within these reefs, place takes on a transformative spatial quality, defined by interconnected power dynamics that flow and shift much like the tides surrounding the Écréhous and Minquiers (Lefebvre 1991, Massey, Allen & Sarre 1999). These dynamics unlock a series of relationships and their social significance at specific moments. This cultural construction reveals how the Écréhous and Minquiers exert a significant impact on human relations and serve as pivotal locations in shaping the allure of islands, as discussed by Baldacchino (2012).

When applying the aquapelagic concept to the Écréhous and Minquiers, we consider not only the relationship between land and sea, but also ideas of island and archipelagic inbetweenness with these reefs because of their peripheral and contested existence between British and French waters, as well as the customary fishing rights that result in nebulous political borders (Baldacchino 2008; Stratford et al. 2011). In this context, the reefs offer a sense of liminality in terms of their construction of place across four distinct domains: fishing, politics, environmentalism and tourism. Such assemblages become contested not only between international borders, but also between different stakeholders with interests on the reefs. As Baldacchino (2008, 220) has commented, "in spite of international borders attempting to be precise, they can at times prove fuzzy and somewhat ambiguous spaces". In this setting, the Écréhous and Minquiers offer examples of contested space, or "integrated terrestrial and marine systems … not necessarily safe or stable entities" (Hayward 2023, 7), which is especially illustrative of the assemblage of aquapelagic place. The methods used in this chapter are drawn primarily from critical interpretative discourse theory, interpreting scholarly, professional and media literature pertaining to the reefs and offering a critical discussion framed within an aquapelagic perspective. Both authors have visited the reefs, for leisure, fishing and tourism, and they have published their research in the field of Island Studies (Bicudo de Castro, Fleury & Johnson 2023; Fleury 2013; Fleury & Johnson 2015; Johnson & Fleury 2017, 2018). This chapter draws on these earlier studies, offering a rethinking of existing literature and extending the discussion specifically to the concept of the aquapelago.

The chapter is divided into four main sections. This first outlines the geographic and political context of the reefs, the second section looks at fishing and traditional rights and the third section explores environmentalism and tourism. The concluding section discusses, in particular, their importance not only for their assemblage of terrestrial and marine environments, but also how other domains are perceived in the making of aquapelagic place.

Geographic and Political Context

The Écréhous and Minquiers reefs are located, respectively, about 9 km northeast and 20 km south of their Jersey mainland. Both reefs extend their respective island parishes,[2] St Martin and Grouville, across the liminality of aquatic space. The reefs are not only a part of Jersey's territory, but also part of its cross-border maritime history, spanning a relatively short distance from Jersey, and slightly farther to France across the Passage de la Déroute (the French name for the body of water between France and the Channel Islands). The reefs are administered by the Bailiwick of Jersey, and "in the present era, … they also occupy a position in shared fishing waters between Jersey and France where fishing, environmentalism and tourism prevail" (Fleury & Johnson 2015, 165). Both reefs are currently uninhabited, but each has some small stone huts (*baraques*), originally constructed in the nineteenth century for quarrymen (as in the case of the Minquiers), then for fishermen, and more recently for leisure (Chambers, Biney & Jeffreys 2016). The huts may sit on the highest rocks, but the size of the reefs should also be measured according to their vastness at low tide.

The geographic configuration of the Écréhous is in a linear shape mainly constituted by a narrow chain of reefs running over less than two kilometers. At low tide, some of the islets are joined by exposed rock, shingle or sand (Fig. 2.1). The largest islet is Maîtr'Île, which has a history of small-scale settlement and the remains of a monastery (Rodwell 1986), although the second largest islet, La Marmotière, has the most extant huts on it. The Minquiers, by contrast, cover an area of about 200 km² at low tide – almost twice the surface of the island of Jersey itself – delimited by cardinal buoys (Coom 2001; Falle 2001; Mallison 2011; States of Jersey Planning and Environment Committee 1999, 251). At high tide, however, there are just nine islets and rocks that remain visible, the largest of which is La Maîtresse Île, which is about 100 m long and 50 m wide and the only islet with huts built on it (Fig. 3.2) (States of Jersey Planning and Environment Committee 1999, 251).[3]

The ebb and flow of the sea, which is about four times over a 24-hour period, changes the nature of the reefs daily. That is, the movement of the sea around Jersey reveals massive areas of sand, shingle and rock as the tide

FIGURE 3.2 The Écréhous at low tide. (Henry Johnson, 2014)

FIGURE 3.3 The Minquiers at mid-tide. View from a sandbank to La Maîtresse Île. (Henry Johnson, 2014)

goes out, but at high tide what is left showing of the reefs appear as mere dots on the water's surface. These semidiurnal tides are characterised by changing and challenging marine conditions due to the combination of two overriding elements: the presence of countless rocks, reefs and sandbanks, which emerge and disappear with tidal flows, and the strength of the currents of tides, which are among the highest in the world (up to 12 m). These two factors combine to make the waters around the reefs particularly hazardous to navigate as rocks become a danger as tides recede. Yet these flows also create a very changeable amphibious and distinct seascape/landscape, part of an aquapelago determined by horizontal and vertical space that changes with tidal flows. Within the ebbs and flows of tides, the reefs are "the British Isles that [almost] disappear everyday" (Stables 2022), and they generate an aquapelagic place. As one hut owner on the Minquiers noted: "The Minquiers plateau is an area of unique, constantly and dynamically changing beauty. It is a place to enjoy vast skies, and multi-toned sea, rocks and sand" (in Fleury & Johnson 2015, 168). As the seascape changes several times over any one day, low tide reveals very large expanses constituted with intertwined sand, rocks and channels, while high tide shows rocks and islets as though they are floating on water. Added to the fact that the situation is never the same depending both on the time and on the tidal coefficient of the day, it is not surprising that only experienced navigators are able to safely visit these areas.

Fishing and Traditional Rights

Located in the Gulf of Saint-Malo, the Channel Islands have long been a disputed place, a source of frequent conflicts and incursions from the early thirteenth century to the end of the war between France and England in 1815.

As part of the Norman Duchy from 933, the Channel Islands remained duti-
ful to their Duke (the King of England), even after continental Normandy
fell to the French in 1204. Forgotten in the clauses of the agreements leading to the reattachment of Normandy (Everard & Holt 2004), the Channel
Islands stayed under pressure from their French mainland because of ongoing tensions between the world's two neighboring superpowers of that era.
The Channel Islands have historically posed a threat to France, serving as a
base for privateers or attempting to influence French politics, including the
French Revolution (La Morandière 1986). Smuggling generated by the border context was facilitated by the labyrinthine configuration of the area and
the proximity of the Écréhous and Minquiers with the neighboring Chausey
Islands (belonging to France) to the east of the Minquiers and just off the
French mainland (McLoughlin 1997).[4]

The unfolding of the international Law of the Sea, combined with local
or customary aspects related to the seas around the Channel Islands, created a complicated history of both social and geopolitical territorialization
of the region. For example, in the 1820s and 1830s, France and the United
Kingdom, which is responsible for the international affairs concerning the
Channel Islands, strived to solve a local conflict known as the "Oyster
Wars", which was triggered by the growing activity of fishers based in Jersey
and traveling the short distance to dredge oysters very close to the French
coast (Gibon 1918). For example, in 1828,

> An unpleasant affair has taken place between the English fishers off the
> coast of Jersey and two French vessels of war, which has led to serious
> consequences, many lives having been lost. The collision has arisen out
> of a question with the French Government, which previously had been a
> subject of serious discussion, concerning the right of fishery on the coast
> of Normandy. About 300 sail of English vessels are engaged in oyster-
> fishing on the coast of Jersey, towards the French shore, and we under-
> stand they have been repeatedly warned not to approach within a certain
> distance of the French coast. These warnings have been little attended to,
> and two French vessels of war, stationed in that quarter, captured and
> took into port an English boat. On this intelligence reaching Jersey, all the
> fishing-smacks proceeded to the French coast, boarded the vessels of war,
> retook the English boat, and brought her back in triumph to Jersey; but
> several of the boatmen have lost their lives, and a considerable number
> were taken prisoners and are in irons.
>
> *(Unattributed, 1828)*

This dispute had resulted in such a number of serious incidents that the
two countries decided to clearly define the rules of fishing. The Law of
the Sea was then at a primary stage, integrating case by case the emerging

norm of the traditional "three-mile limit", which is nowadays mostly out-dated. This limit was included in the Granville Bay Agreement, signed in 1839 between France and the United Kingdom in which we can find: (a) a three-mile band around Jersey that was forbidden to French fishers, (b) an irregular border called the "A to K Line" to the east of the Écréhous, to which Jersey fishers had no access, and (c) between them a large space called Common Sea accessible to both parties, in which the Écréhous and Minquiers are located. This was a very early agreement, one of the first – if not the first – of its kind (Labrecque 1998), and it seemed to work to eve-rybody except that an increasing number of voices in Jersey wished to get international acknowledgment of the Bailiwick's sovereignty over the two reefs. In the absence of mutual agreement, the United Kingdom and France decided in 1951 to submit litigation at the International Court of Justice in The Hague, which was resolved in Jersey's favor in 1953. However, a prior agreement had been reached in 1951 between the parties, which specified that the losing party in The Hague would continue to be able to fish in the waters of the winning side. This agreement particularly concerned the Écréhous and Minquiers, which were highly valued because of their natural marine resources, especially shellfish. Both parties came to The Hague with very different political positions. Jersey mainly put forward the fact that the huts on both reefs had been built by residents of Jersey whose administra-tive services had registered the title deeds and compiled various other legal documentation (Bicudo de Castro, Fleury & Johnson 2023; Mallison 2011, 73; Wright 1897, 96). The French position was based primarily on custom-ary fishing rights.

The huts on the Minquiers were not known in 1748 (Godfray 1929, 194), although toward the end of the eighteenth century there would have been some form of dwelling(s) to accommodate the quarrymen who were based there to quarry granite for the building of Fort Regent on Jersey. By 1903, however, there were 18 huts on the main islet, which were mostly built of stone (Mallison 2011, 73). The exact number has fluctuated over the years, with 16 dwellings shown on a map of 1929 (Godfray 1929, 192), and with just 12 remaining today. Two of these huts are owned by the States of Jersey as Crown property and are used by Jersey Harbours and Customs. In a move that brought a new form of transport to the Minquiers, in the 1970s a helicopter pad was built (Falle 2001, 99). On the Écréhous, while La Marmotière is distinct as an islet with 21 small dwellings tightly packed in (States of Jersey Planning and Environment Committee, 1999, 243), there are a few nearby rocks connected by a one-kilometer-long shingle bank (La Taille) that have several huts built on them. The huts may offer a visual refer-ence to settlement, but settlement in these settings is based mostly on short-term visits, with the huts acting as symbols of local proprietary, and they are highly sought after buildings.

At The Hague, the French side argued on the very ancient and continued presence of its fishers mainly coming from the neighboring archipelago of Chausey to the southeast of the Minquiers, as well as the fact France had progressively secured the dangerous sea access to the ports of Granville and Saint-Malo with beacons, including to the west and south. It was not surprising that the British arguments convinced the Court, and on 17 November 1953 the Écréhous and Minquiers were officially recognized as belonging to the British Crown, as part of the Bailiwick of Jersey (Johnson 1954, 209).

The decision of the International Court of Justice was a prerequisite for further negotiations between France and the United Kingdom to clarify the border between France and the maritime sphere of the Bailiwick of Jersey. But it was not before 1987 that the process began. After 13 years of negotiations "rock by rock",[5] a bilateral agreement establishing both the international boundary between Jersey and France and mutual fishing access was signed in Jersey on 4 July 2000. The right for islands to benefit from a territorial sea had since been extended from 3 to 12 nautical miles,[6] but because the distance between the two countries is less than 12 nautical miles, the rule of equidistance would normally be adopted. However, with the Écréhous and Minquiers, while they are at the edge of Jersey's physical territory, they are not permanently inhabited so the border between these reefs and France was not equidistant (Dobelle 2000).

The Granville Bay Agreement signed in 2000 has two parts. The delimitation of the border was strictly technical, but another challenge was to articulate this intangible line with the need to consider the respective historical and customary fishing rights (Fleury 2011). A Joint Advisory Committee of the Bay of Granville had the task to give consistency to the treaty by drawing up provisions for clarifying fishing regulation. Their composition, competences and functioning are described within a document attached to the international treaty that defines the boundary. There were three parties – Jersey, Normandy and Brittany – which met three times a year, respectively, in St. Helier, Granville and Saint-Malo. The meetings had three types of participants, the professionals (fishermen as well as their representatives), the representatives of both administrations (France and Jersey) and biologists from both nations.

The implementation of cross-border fishing was put forward by French fishers as an essential point. The result has been a model of detailed zoning with several areas accessible based on quantity of catch, port of origin and length and power of the fishing vessels. The Écréhous and Minquiers were in the Common Sea (i.e., common to both countries), between these limited zones and their access did not suffer from restrictions other than the administering of licenses from the respective authorities based on recognizing historical fishing practices. Regarding the Minquiers, a functional distinction should be made between their outer limits – but inside the buoys – where dredging and trawling are possible and their inner areas where the complexity of navigation limits other fishing activities to areas, such as potting as practiced in the area only by a handful of fishers from France and Jersey.[7]

The functionality of this agreement remained intact until the disruptive event of Brexit in June 2016.[8] Although there was not an obvious and immediate connection between Brexit and the Bay of Granville Agreement, a tense atmosphere emerged during the institutional meetings that gathered French and Jersey professionals. In autumn 2018, the Jersey Fishermen's Association decided to boycott the meeting, arguing that the French side had delayed updating the list of authorized boats. A French reluctance to accept the extension of a forbidden zone of dredging in the Écréhous also contributed to the escalation of the situation.

In December 2020, at the end of a lengthy process leading to the effective implementation of the United Kingdom leaving the European Union (EU), a vote in the Houses of Parliament in London approved the integration of the three Crown Dependencies (Jersey, Guernsey and the Isle of Man) in the Trade and Cooperation Agreement (TCA) between the United Kingdom and the EU. This decision was then approved by their respective representative Assemblies on 27 December. This legal process annulled all other provisions and dealt a deadly blow to the Bay of Granville Agreement. What followed was a confusing and sometimes confrontational period between French and Jersey fishers and their representative and governmental bodies (Schatz 2019). For instance, in a context where about 350 French vessels could fish in Jersey waters before Brexit (Taylor 2021, January 20), and now with fishing licenses issued by Jersey based on proven prior fishing activities, Jersey's main newspaper, *Jersey Evening Post*, covered the tensions in much detail, with headlines including: "Fishermen 'wrong to support UK over new law'" (Newsdesk 2020), "Jersey Fishermen facing 'threats and intimidation'" (Taylor 2021, January 20), "Confrontation on the horizon in Fishing row" (Taylor 2021, October 29), and "French Fishermen Fined for illegal Fishing in Jersey waters" (Newsdesk 2022).

While discussions involving Jersey, France, the United Kingdom and the EU took place in 2021, the implementation of the TCA was not really called into question, which resulted in a significant drop in French fishing rights in Jersey's waters. For the Écréhous and Minquiers, given the provisions of the TCA,[9] most of the French pot vessels were acknowledged, although this was not necessarily the case with regard to other fishing activities, which created bitterness on the French side.[10] By getting the exclusive right to deliver licenses to French vessels, which was before under the administration of the flag state, we can consider that, owing to Brexit, Jersey achieved total sovereignty of its territorial waters, including those of and around the Écréhous and Minquiers. Here, the aquapelago around each reef is an assemblage of socially constructed place where inter-state tensions and the activities of fishers reveal political relations between states as drivers for the changing dimensions of maritime space. However, within this space there are other culturally constructed dimensions of the aquapelagos, including environmental and touristic concerns as outlined in the following section.

Environmentalism and Tourism

Over geological time, Jersey's maritime landscape has changed considerably. This is particularly evident in connection with the Écréhous and the Le Ruau underwater channel between Jersey and the reef:

> The Le Ruau channel is a palaeochannel, which – at a time of much lower sea levels – would have been a river channel flowing from mainland France into the sea. At that time, Les Écréhous would have been part of the French landmass.
>
> *(Fiona Fyfe Associates 2020, 152)*

The divide between shallow sea and the deep sea (above 30m deep) to the west of Jersey offers a submarine perspective of Jersey's aquapelago. This is the deepest part of Jersey's aquatic space, "and its bathymetry slopes gradually down towards the west, reaching 50m depth at the Bailiwick boundary" (Fiona Fyfe Associates 2020, 158). This part of Jersey generally has stronger currents than other parts, less temperature variation and less underwater light, therefore offering a very different marine environment to the shallower waters around Jersey's main reefs.

In the modern era, the geological and marine setting of the Écréhous and Minquiers adds to the environmental and touristic capital of Jersey regarding their marine mammals, bird life, fish, crustaceans, plants, sandbanks, shingle beds, rocks, islets and relative remoteness from their Jersey mainland, all of which are frequently featured in reports and other documents about the distinctive biodiversity and geodiversity of these habitats. For example, in the *Jersey Integrated Landscape and Seascape Character Assessment* (see Fiona Fyfe Associates 2020), such surface, intertidal and submarine features are aptly described as the "Jewel in Jersey's crown" (Fiona Fyfe Associates 2020, 132). Such features include a landscape and marinescape of intertidal zones with "rocks, sandbanks, maerl beds, rock platforms, flooded gully complexes, kelp forests and seagrass beds" (Fiona Fyfe Associates 2020, 132).

Both reefs, along with other aspects of Jersey's natural environment, have received international recognition. They are designated conservation areas with both listed in 2005 as Ramsar sites of international importance (States of Jersey 2011, 2012a, 2012b). Both reefs are also Marine Protected Areas and part of Jersey's Coastal National Park (Fiona Fyfe Associates 2020,

136–137). These areas cover the subtidal, intertidal, littoral and terrestrial spheres of the relative aquapelagos, offering a distinct environment over land and sea that "creates a deep sense of remoteness, wildness, tranquillity and detachment" (Fiona Fyfe Associates 2020, 133). The aquapelagos offer an assemblage of marine and terrestrial space that ebbs and flows horizontally and vertically daily and with animate and inanimate markers of environmental capital. Such capital extends to the commercial marketing sphere of environmental and adventure tourism, which have developed rapidly in the twenty-first century and especially offer day trippers an opportunity to visit either of the reefs by high-speed RIB (rigid inflatable boat). Several companies operate such excursions to the reefs, which would typically involve a half-day trip. These companies, Island RIB Voyages[11] and Jersey Seafaris,[12] operate from Jersey, while from France commercialization is primarily in the form of larger sailing yachts. For example, one Jersey operator highlights the trip to the Minquiers as "one of wonders of the natural world", describing the setting as a "paradise on our doorstep" (Island RIB Voyages 2023). The same operator markets the Écréhous in much the same way, emphasizing that with such a trip visitors can "experience the wildlife and wonder the jewel of Jersey has to offer". However, while such operators have brought the reefs closer to both Jersey and France in terms of speed of access and an increased number of visitors, such travel comes with concerns. That is, "increased visitor pressures are a threat" (States of Jersey Planning and Environment Committee 1999, 240). While the reefs' relative geographic isolation from their Jersey mainland is an aspect that is attractive to visitors, that isolation is threatened by the increasing numbers who want to visit. As one Jersey authority notes, "the special sense of remoteness and wideness is being eroded by increasing human disturbance" (States of Jersey Planning and Environment Committee 1999, 240).

With increasing concerns about the number of visitors to the Écréhous, those who have huts on one of the reefs formed the Les Ecrehous' Residents Association in 2009, and they outlined its aims in order:

1. to maintain and preserve the peaceful atmosphere, beauty and tranquility of the Ecrehous;
2. to preserve the natural environment and wildlife of the Ecrehous for all;
3. to ensure the sustainable use of the Ecrehous as a safe and valued environment for the enjoyment of all those who visit the Ecrehous; and
4. to express an opinion or make representations on matters relating to the Ecrehous ('Les Ecrehous' Residents Association 2015).

On the Minquiers, typical complaints about the increased number of day trippers include speeding vessels and jet skis. For example, there have been up to 20 vessels reported on some days at the main anchorage (States of Jersey

2012b, 13). The situation at the Écréhous is more intense and "the sheer volume of general visitors to the reefs can be a potential threat if appropriate management steps are not taken. Up to 80 vessels have been reported on some days with several hundred people on land" (States of Jersey 2012a, 14). Personal communication with one key informant with a professional connection with the reefs even noted differences between the type of visitor to each reef: "There is a camaraderie amongst Minquiers visitors that one does not get at Les Écréhous. This is probably because you need to be reasonably competent to get a boat there and so the reef does not attract quite so many party goers (although it does have them)" (Anon 2015, personal communication). There are also sporting visitors, who are especially from the kayaking community and travel from Jersey. One Jersey-based company that specializes in sea kayaking is Jersey Kayak Adventures, which was established in 2003.[13] As well as private kayakers visiting the reefs, Jersey Kayak Adventures offers guided tours, usually comprising small groups of up to 10 kayakers who not only visit the reefs, but also have the chance of seeing dolphins and other sea and birdlife during the excursion. For the company's tour to the Minquiers, it portrays the reef as a space that is quite different from the local Jersey waters. As it notes: "Kayak into a world of huge sandbars surrounded by Caribbean blue seas and rugged rocky outcrops" (Jersey Kayak Adventures 2023).

In this contested aquapelagic context – and moving from the reefs as a place for fishers – both the Écréhous and Minquiers have become locations that offer a site of ownership for Jersey residents that would require travel by sea to reach them. Such locations can provide only a limited number of dwellings, thereby increasing the value of ownership, and particularly for those who would also have a boat to reach their property. While a retreat for owners of property on the reefs, as boat owners, such visitors would need to be experienced on the sea to navigate the treacherous waters around the reefs, each being surrounded by submerged rocks and fast moving tidal flows. The reefs, therefore, help create layers of aquapelagic place from maritime and terrestrial space, often contested and ever changing in a setting of increasing anthropogenic disturbance.

Conclusion

Jersey has various aquapelagic regions. While one overarching aquapelagic area aligns with the jurisdictional territory as defined on a political map of both land and sea within the Bailiwick, there exist additional aquapelagic domains within Jersey's maritime boundaries that intersect with French territory. These additional regions are shaped by the intricacies of fishing politics and experienced within environmental and touristic spheres. This chapter has highlighted the Écréhous and Minquiers as examples of such aquapelagic regions, representing a fusion of land and sea, and characterized by nebulous and changing boundaries, both horizontally and vertically, dictated by tidal flows and the politics of place. The contemporary significance of these reefs as markers of environmental and touristic value becomes evident within this intricate political landscape. Here, the natural surroundings and environment create a transitional space – an in-between, liminal zone – on the edge of political borders and attracting competing interests. This phenomenon not only extends to the fishing community, but also encompasses environmental policies that position the Écréhous and Minquiers as symbols of a unique natural world. It also extends to the realm of tourism, where the seascapes and landscapes become symbols of authenticity and allure, contributing to the concept of aquapelagic capital. Not only is there a "lure of the island" (Baldacchino 2012), but also a "lure of the aquapelago". Their study has shown further ways of interpreting and experiencing two unique settings where marine and terrestrial space becomes an important place in human life. The Écréhous and Minquiers have been examined through the perspective of an aquapelagic lens. This chapter has shown not only the coexistence of land and sea, but also the dynamic interplay of these elements, constantly shifting with the tides. These natural rhythms contribute to the cultural value of these locations, enriching their importance in the fishing industry as well as enhancing their appeal to environmental politics and tourists alike. Through this exploration, we have shown a fresh avenue for interpreting these two distinctive settings, from political, environmental and touristic standpoints. In these spaces, while not permanently populated, where the boundaries between marine and terrestrial realms blur, the aquapelago becomes integral to human life and offers unique opportunities for engagement with the natural world.

Notes

1 In the Channel Islands, there is also the Bailiwick of Guernsey, which includes the island jurisdictions of Guernsey, Alderney and Sark, along with several smaller islands, islets, reefs and rocks.
2 There are 12 parishes on Jersey (on Jersey's parishes, see Hargreaves 2023).
3 Based on a detailed map displayed by Chambers, Biney and Jeffreys (2016, 142), one can estimate that Maîtresse Île covers an average of about two hectares at low tide and one hectare at high tide.
4 The Chausey Islands have themselves at times been contested territory between England and France (Johnson 1954, 198–199).
5 According to the words of Simon Bossy, one of Jersey's key negotiators (Chambers, Biney & Jeffreys, 2016, 116).
6 Jersey applied this principle through its *Territorial Sea Act* (1997); see https://www.jerseylaw.je/laws/current/Pages/15.800.aspx.
7 Regular trips in the Minquiers on a fishing boat by the first-named author as well as personal communication with professionals make it possible to estimate at less than ten the number of pot vessels either from France or from Jersey

regularly fishing lobsters in the inner part of the Minquiers. The assertions in this part of the discussion derive from the same author who attended the Bay of Granville meetings from 2000 to 2018.

8 Brexit refers to the process by which the United Kingdom decided to leave the European Union, culminating in its formal withdrawal on 31 January 2020, following a referendum held on 23 June 2016 where a majority of UK citizens voted in favor of leaving the EU.

9 French traditional rights were maintained for vessels demonstrating activity in Jersey waters of at least 11 days during a period of 12 months ending on 31 January between 1 February 2017 and 30 January 2020.

10 The absence of the possibility of being issued a license from Jersey for a young fisherman and the problem of replacing vessels raise fears of an eventual extinction of French historic rights in Jersey waters.

11 See http://www.jerseyribtrips.com/ (accessed 25 September 2023).

12 See http://www.jerseyseafaris.com/ (accessed 25 September 2023).

13 See http://www.jerseykayakadventures.co.uk/ (accessed 25 September 2023).

References

Baldacchino, Godfrey. 2008. Islands in between: Martín García and other geopolitical flashpoints. *Island Studies Journal*, 3(2), 211–224.

Baldacchino, Godfrey. 2012. The lure of the Island: A spatial analysis of power relations. *Journal of Marine and Island Cultures*, 1(2), 55–62.

Bicudo de Castro, Vincent, Fleury, Christian and Johnson, Henry. 2023. Micronational claims and sovereignty in the Minquiers and Écréhous. *Small States & Territories*, 6(1), 35–48.

Chambers, Pauk, Biney, Francis and Jeffreys, Gareth. 2016. *Les Minquiers: A natural history*. Totnes: Charonia Media.

Coom, David. 2001. Livings from the reef. In Anthony Rowland, ed. *Grouville, Jersey: The history of a country parish*. Grouville: Parish of Grouville, 100–105.

Dobelle, J.-F. 2000. Les accords franco-britanniques relatifs à la Baie de Granville du 4 juillet 2000. *Annuaire Français de Droit International*, 46, 524–547.

Everard, J.A. and Holt, J.C. 2004. *Jersey 1204: The forging of an Island community*. London: Thames & Hudson.

Falle, Richard. 2001. Les Minquiers. In Anthony Rowland, ed. *Grouville, Jersey: The history of a country parish* (pp. 91–99). Grouville, Jersey: Parish of Grouville.

Fiona Fyfe Associates. 2020. *Jersey integrated landscape and seascape character assessment*. St Helier: Strategic Policy, Planning and Performance, Government of Jersey.

Fleury, Christian. 2011. Jersey and Guernsey: Two distinct approaches to cross-border fishery management. *Shima*, 5(1), 24–43.

Fleury, Christian. 2013. The island/sea/territory relationship: Towards a broader and threedimensional view of the aquapelagic assemblage. *Shima*, 7(1), 1–13.

Fleury, Christian and Johnson, Henry. 2015. The Minquiers and Écréhous in spatial context: Contemporary issues and cross perspectives on border islands, reefs and rocks. *Island Studies Journal*, 10(2), 163–180.

Gibon, Paul De. 1918. *Un archipel normand: Les Îles Chausey et leur histoire*. Coutances: Notre-Dame.

Godfray, A. D. B. 1929. Archaeological researches at the Minquiers. *Société Jersiaise Annual Bulletin*, 11(2), 193–199.

Hargreaves, Peter. 2023. Jersey Parishes, iconography and Island senses of place. *Shima*, 17(1), 91–122.

Hayward, Philip. 2012a. Aquapelagos and aquapelagic assemblages. *Shima*, 6(1), 1–11.

Hayward, Philip. 2012b. The constitution of assemblages and the aquapelagality of Haida Gwaii. *Shima*, 6(2), 1–14.

Hayward, Philip. 2023. Extraordinarily hazardous: Fog, water, ice and human precarity in the aquapelagic assemblage of the Grand Banks. *Coolabah*, 34, 7–24.

Island RIB Voyages. 2023. https://www.islandribvoyages.je/voyages/minquiers

Jersey Kayak Adventures. 2023. https://jerseykayakadventures.co.uk/les-ecrehous -les-minquiers-sark-offshore-trips/

Johnson, D. H. N. 1954. The Minquiers and Ecrehos case. *The International and Comparative Law Quarterly*, 3(2), 189–216.

Johnson, Henry and Fleury, Christian. 2017. The Minquiers and Écréhous, Channel Islands - Single sovereignty but shared jurisdiction. In Godfrey Baldacchino, ed. *Solution protocols to festering island disputes: 'Win-win' solutions for the Diaoyu/Senkaku Islands*. New York: Routledge, 103–108.

Johnson, Henry and Fleury Christian. 2018. Tourism and resilience on Jersey: Culture, environment, and sea. In J. M. Cheer and A. A. Lew, eds. *Tourism, resilience and sustainability: Adapting to social, political and economic change*. New York: Routledge, 85–102.

Labrecque, Georges. 1998. *Les frontières maritimes internationales: Essai de classification pour un tour du monde géopolitique*. Paris: L'Harmattan.

La Morandière, Charles de. 1986. *Histoire de Granville*. Granville: Roquet.

Lefebvre, Henri. 1991. *The production of space*. Translated by D. Nicholson-Smith. Oxford: Blackwell.

Les Ecrehous' Residents Association. 2015. http://www.ecrehous.com/index.html

Mallison, Jeremy. 2011. *Les Minquiers: Jersey's southern outpost*. Bradford on Avon: Seaflower.

Massey, Doreen, Allen, John and Sarre, Philip, eds. 1999. *Human geography today*. Cambridge: Polity Press.

McLoughlin, Roy. 1997. *Sea was their fortune: A maritime history of the Channel Islands*. Bradford on Avon: Seaflower.

Newsdesk. 2022. September 29. French Fishermen Fined for illegal Fishing in Jersey waters. *Jersey Evening Post*. https://jerseyeveningpost.com/news/2022/09/29/ french-fishermen-fined-for-illegal-fishing-in-jersey-waters/

Newsdesk. 2020. October 20. Fishermen 'wrong to support UK over new law.' *Jersey Evening Post*. https://jerseyeveningpost.com/news/2020/10/20/fishermen -wrong-to-support-uk-over-new-law/

Rodwell, Warwick, J. 1986. Saint Mary's Priory, Les Écréhous, Jersey: A reappraisal. *Société Jersiaise Annual Bulletin*, 24(2), 225–231.

Schatz, Valentin. 2019. Access to fisheries in the United Kingdom's territorial sea after its withdrawal from the European Union: A European and international law perspective. *Goettingen Journal of International Law*, 9(3), 457–500.

Stables, Daniel. 2022. August 29. The British islands that disappear every day. *BBC travel*. https://www.bbc.com/travel/article/20220828-the-british-islands-that -disappear-every-day

States of Jersey Planning and Environment Committee. 1999. *Jersey Island Plan review: Countryside character appraisal. Offshore reefs and islands*. St Helier, Jersey, States of Jersey Planning and Environment Committee.

States of Jersey. 2011. *Jersey's south east coast Ramsar Management Plan*. Trinity, Jersey: Department of Planning and Environment.

States of Jersey. 2012a. *Les Écréhous and Les Dirouilles Ramsar Management Plan*. Trinity, Jersey: Department of the Environment.

States of Jersey. 2012b. *Les Minquiers Ramsar Management Plan*. Trinity, Jersey: Department of the Environment.

States of Jersey. 2024. Jersey's relationship with the UK and EU. https://www.gov.je /government/departments/jerseyworld/pages/relationshipeuanduk.aspx

Stratford, Elaine, Baldacchino, Godfrey, McMahon, Elizabeth, Farbotko, Caroline and Harwood, Andrew. 2011. Envisioning the archipelago. *Island Studies Journal*, 6(2), 113–130.

Suwa, Jun'ichiro. 2012. Shima and aquapelagic assemblages: A commentary from Japan. *Shima*, 6(1), 12–16.

Taylor, Edward. 2021, January 20. Jersey Fishermen facing 'threats and intimidation'. *Jersey Evening Post*. https://jerseyeveningpost.com/news/2021/01/20/jersey -fishermen-facing-threats-and-intimidation/

Taylor, Edward. 2021, January 28. New Fishing agreement relaxes French tensions. *Jersey Evening Post*. https://jerseyeveningpost.com/news/2021/01/28/new -fishing-agreement-relaxes-french-tensions/

Taylor, Edward. 2021, October 29. Confrontation on the horizon in Fishing row. *Jersey Evening Post*. https://jerseyeveningpost.com/news/2021/10/29/ confrontation-on-the-horizon-in-fishing-row/

Territorial Sea Act 1987 (Jersey) Order 1997. Privy Council of Jersey. https://www .jerseylaw.je/laws/current/Pages/15.800.aspx

Unattributed. 1828. In the course of a few days. *The Times*. https://link-gale-com .ezproxy.otago.ac.nz/apps/doc/CS35018357/TTDA?u=otago&sid=bookmark -TTDA&xid=908bc92e

Wright, Margaret Baker. 1897. *Hired furnished: Being certain economical housekeeping adventures in England*. Boston: Roberts Brothers.

4

THE PRECARIOUS AQUAPELAGIC ASSEMBLAGE OF THE GRAND BANKS (NORTHWEST ATLANTIC)

Philip Hayward

Introduction

This chapter emphasizes the manner in which aquapelagos can constitute highly precarious assemblages for humans to negotiate. In particular, it examines the disruptive role that fog, associated weather and sea conditions play in livelihood activities undertaken on and around the Grand Banks of the northwestern Atlantic, the affective atmosphere they create and their effect on human participants. After an introduction to the position and nature of the Grand Banks, relevant weather systems, ocean currents and iceberg trajectories through the region, the chapter profiles the nature of fishing (and, subsequently, oil extraction) in the area, the precarity of fishery activities and their reflection and inscription in various media. Within this, the case study examines the manner in which fog is not so much an uncomfortable intrusion into an otherwise manageable industrial operation as a key characteristic to be accommodated. This emphasizes the manner in which human experience of fog is central to the socioeconomic experience of the Grand Banks as an aquapelagic space that is constituted around them, extending the aquatic into the atmospheric.

The Grand Banks: Position and Features

The Grand Banks is an elevated, approximately 93,200 square kilometre area of continental shelf that peaks at between 15 and 90 metres below the ocean surface approximately 330 kilometres southeast of the southern tip of Newfoundland (Fig. 4.1). The area is located at the point that the southerly flowing, cold Labrador Current first abuts against and subsequently flows

DOI: 10.4324/9781003569534-4

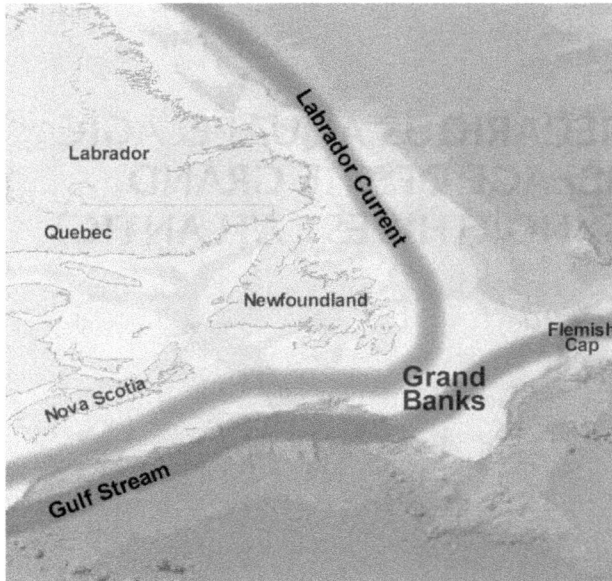

FIGURE 4.1 Map of the northwest Atlantic shelf showing positions of Newfoundland and the Grand Banks and the Labrador and Gulf Stream Currents. (Wikicommons, 2006 UTC)

north of the warm, northeasterly flowing Gulf Stream. The contact of the moist, warm air occurring over the latter and the cold surface of the former causes advection (the horizontal transfer of heat), which produces concentrated water vapour that manifests as fog. The continuous nature of the two currents results in advection fog forming for the majority of the year (often reported as around 300 days p.a.), most predictably and intensely in mid-Spring to mid-Summer , when fogs are often prevalent for extended durations. As a result, the Grand Banks are commonly recognized as the most fog-bound area of the planet, an attribute that has caused major logistical difficulties for shipping, fishers and the oil industry. The reliable presence of fog in the region in mid-Spring to mid-Summer, even during windy periods,[1] coincides with the regular arrival of sea ice and icebergs in the area, swept down from Greenland and the Canadian Arctic on the Labrador Current. As the passengers and crew of the RMS *Titanic* found in the southern area of the Grand Banks on April 14, 1912, the encounter of icebergs and ships can often be catastrophic for the latter. If this combination of phenomena were not hazardous enough, the Grand Banks is also commonly acknowledged as one of the most storm-prone areas of the planet. As the US Department of Homeland Security Navigation Center of Excellence (USDHSNCE) (n.d.) has detailed:

The tracks of the storms leaving the North American continent frequently pass over or near the Grand Banks, particularly in the winter and early spring. In some cases the storms explosively intensify as they leave the east coast, and by the time they reach the Grand Banks they carry storm or hurricane force winds. These storms are sometimes called "weather bombs", which is defined as a low pressure system that experiences a decrease in central pressure of 24 millibars in 24 hours. Winds can exceed 100 km/h (62 mph) and seas 15 m (about 50 ft).

As the USDHSNCE has summarised, with little (if any) exaggeration, the "combination of frequent storms, persistent fog, and the presence of sea ice and icebergs creates *extraordinarily hazardous* conditions for mariners" (my emphases).[2] Given the extreme precarity of maritime operations in the area, at *any* time of year, and particularly in Spring and Summer, it is pertinent to consider why maritime livelihood activity has developed and persisted in this region. The answer is relatively straightforward. The mixture of warm and cold waters around the Banks enables nutrients to be constantly elevated from the shallows, ensuring an abundance of fish – principally cod, hake and haddock – in the area before they were catastrophically over-fished in the mid- to late 20th century. The high value of this product in global markets over an extended duration has ensured that there have been substantial profits to be made, however precarious the livelihoods of those involved.

In the absence of any evidence to the contrary, it seems unlikely that the Grand Banks were fished by indigenous communities from (present-day) Newfoundland, the Canadian maritime provinces or the US New England states (the closest land areas to the Banks) in the period prior to European intrusion into the northwest Atlantic in the late Middle Ages. This is not to say that the maritime technologies, skills and/or collective enterprise required were beyond the capabilities of the Beothuk or Mi'kmaq peoples of the region, but, rather, that there is no evidence that these were galvanised in pursuit of the fishery, perhaps due to the plentiful supply of fish in safer, more immediate offshore waters (Gilbert 2011), thereby ensuring that the Grand Bank's fish biota remained unexploited by humans until Western fishing commenced in the area in the late 1500s.

Holm, Ludlow, Schere et al. (2018, 1) have identified that the discovery of the Grand Banks and its recognition as a "super-abundant" fishing ground (for cod, in particular) by European mariners in the late 1400s and early 1500s precipitated a "fishing revolution" in Europe. This occurred because fish was "a high priced, limited resource" due to "an already-severe depletion of European waters" (Holm, Ludlow, Schere et al. 2018, 14). The authors have also contended that the exploitation of the Grand Banks by fishers, initially from the Basque region of Spain and, subsequently, from England and France, led to "a 15-fold increase of cod … catch volumes and likely a tripling of fish protein to the European market" (2018 , 1). Given that the European market was geared to fresh fish supply, wholesale and retail, the extreme distance of the Grand Banks fishing grounds from Europe (c. 3,750 kilometres) mitigated against cod being delivered in sufficiently fresh form (given the lack of ice production and/or refrigeration in this period) and required the fish to be processed, in this case, dried and salted. The small size of fishing vessels and the difficulty in establishing stable drying and salting environments on ships at sea in stormy waters meant that the processing enterprise was best undertaken on land, and, particularly, on the nearest land to the Banks: Newfoundland's southern shore. Seasonal drying operations led to individuals or small groups of mariners "wintering over" to look after drying facilities, thereby creating the first all-year round (micro-) colonies of Europeans on the island.

This pattern of resource exploitation created an aquapelago – a socially constructed space derived from island or coastal locations where humans have developed particularly concentrated engagements with the marine environment for their livelihoods. Evidence of the perception and constitution of this space has been outlined by Rankin and Holm in their (2019) study of the early maritime cartography of the region. Such studies illustrate, as Maxwell (2012) has ably identified, that there are *places in the sea* that merit attention as assemblages in their own right and that can be inscribed – in various ways, in various media – in a manner that "sets subjective and objective epistemologies into productive dialogue" (Maxwell 2012, 23).

From the early 18th century on, as European colonial fishing communities became established in the northwest Atlantic, the central area of the aquapelago – i.e. the Grand Banks – was linked through separate but parallel industrial enterprises to communities on the shores of Newfoundland, New England and the French micro-colony of St Pierre and Miquelon and on to major markets in western Europe, the northeast coast of America and European colonies established in the West Indies.[3] While the broad settlement of New England derived from various factors, Newfoundland and St Pierre and Miquelon were founded as terrestrial adjuncts of the maritime area of the Banks, as exploited by European powers, and, thereby, the livelihood activities on the Banks preceded and prefigured the establishment of the land colonies.

The decline of the Grand Banks fishery and the near-total collapse of cod stocks in the 1990s, following peak, unsustainable extraction in the 1970s, is well known[4] and marks the (current) endpoint[5] in the Banks' active maintenance as the focus of Newfoundlandic and transatlantic aquapelagic systems focussed on cod extraction. This illustrates a key point about aquapelagos present in early discourse about them but developed more thoroughly by subsequent discussants (e.g. Guerin 2019), namely that, as entities constituted by human presence in and utilisation of the environment – they come into being at particular moments and change or decline due to environmental factors and/or as socioeconomic enterprises, technologies, and/or the resources and trade systems they rely on develop or decline. If aquapelagos are "performed entities", they are ones that also involve sets of subsidiary performances that involve humans interacting with various aquapelagic actants. Enter the fog…

Fog on the Banks/ Humans in the Fog

While the Grand Banks was an "extraordinarily hazardous" area to fish from its earliest stages, with regular losses of ships, an industrial innovation raised the precarity of the enterprise for the humans involved in hauling the catch from the icy waters of the region. Until the early 1800s, fishing was undertaken by crewmembers casting and hauling in handlines from the decks of schooners. While arduous, the schooners' decks and gunwales offered some stability, a degree of protection from being washed overboard and some shelter from the harshest aspects of the elements. However, initial experiments with launching dories (small boats with high, flared sides) from schooners showed that handline fishers working in these over a dispersed area could haul in greater volumes of fish than the same number could working alongside each other on a schooner deck. As a result, by the 1860s dory fishing was the dominant industrial practice (Stoodley 2021). It was also a far riskier one. The risks were multiple and included being swamped in high seas and the regular danger caused by fog.

Fog is particularly disruptive to human perception and navigation capabilities. Dense concentrations of water vapour foreshorten the limits of human visual perception and also impair our abilities to distinguish objects and their proximity through diminishing – "greying out" – the stimulus contrast (the ratio of foreground to background luminance) between dark and light objects, thereby obscuring details. As Harley, Dillon and Loftus (2004, 198) have identified, "although there are many ways in which visual stimuli may be degraded, a large body of research indicates that stimulus contrast… is critical in determining the fundamental response of the visual system." Their study also suggests that impairments to distinguishing visual stimuli might complicate mental processing of such stimuli and, consequently, adversely impact recognition, memory and related cognitive operations:

> The data from all the experiments allow the conclusion that some function of stimulus contrast combines multiplicatively with stimulus duration at a stage prior to that at which the nature of the stimulus and the reason for processing it are determined, and it is the result of this multiplicative combination that determines eventual memory performance.
>
> *(Harley, Dillon & Loftus 2004, 198)*

Put more simply, low-contrast visual environments can be confusing and impressions gained from them can be unreliable. Persistent fog thereby impairs human abilities to navigate through visual orientation (which relies on a positional sense that utilises memory) and/or to track their trajectories through space.

Acknowledging fog's complication of standard perceptions, Martin (2011) has focussed on the human experience of being in fog and of the bodily apprehension of it, asserting that fog acts as a "gathering-force, intensifying the immanent entanglements of body *with* world" that give fog – and human experience of fog – an associative dimension:

> Through its opacity fog connotes feelings of density and dislocation, the suspension of water droplets mid-air communicating a sense of inaction, of stillness. ... Alongside such representations the presence of fog also forces a phenomenological engagement with embodied immersion: one's body enveloped and entwined with space.
>
> *(Martin 2011, 454)*

Supporting this characterisation, Martin quotes Michel Serres's (2008, 69) contention that "Night is empty or hollow, fog is full; darkness is ethereal, mist is gaseous, fluid, liquid, viscous, sticky, almost solid". These perspectives accord with more general work that has been undertaken on what has been termed the "atmospheric turn" in anthropology (see Griffero 2019) and, more specifically, to critical work addressed to perceptions of space and of being in space. Vale's work on the mistiness of the Azores (2018) is a striking example of the latter. She asserts that, "bodily apprehension-immersion" and processes in which "the subject recurrently and continually receives impressions" are crucial to experiences and representations of the islands (Vale 2018, 92). More broadly, in a contention that could serve as a summary of the framework for this chapter, Hodges (2022, 122) contends that the atmospheres – understood as affective dimensions of places:

> are more than just isolated destinations, material resources or idyllic representations. They are ecosystems, spheres of influence and sites of collision between competing systems: of the colonial and indigenous, the geologic and oceanic, the gaseous and the liquid. [And to] study these environs involves thinking about their relationship to wider networks and assemblages, both real and imagined.

Returning to fog with such a conceptual framework, Martin identifies that his discussion derives from considering the logistics of movement-space and of how fog disrupts this and needs to be accommodated within such logistics. He moves beyond a discussion of foggy atmospheres to argue that:

fog positions us in the midst of the aerial, intensifying the schism between lightness and heaviness, entangling us with space in a determinedly material sense. By highlighting such a problematic of immersion, I suggest that fog momentarily strands us in the instant of disorientation, leaving us to ponder our conditional engagements with the near and far.

(Martin 2011, 454)

Drawing on his own experience of walking along the English coast in fog, he reflects that:

the fog was not simply disorienting in its thickened presence, it also gathered together the spatial, illustrating in an acute sense the relational tenacity of space. Similarly, the corporeal relationship to this phenomenon was such that the bodily inhabitation of space was intensified.

(Martin 2011, 455)

Shifting from Martin's focus on the aerial geography of the English coast to the present chapter's address to aquapelagic experience on the Grand Banks, we can regard the latter in terms of an "admixture" of ocean, air, ship and ice. Martin's discussions of the physical sensation of fog – which are echoed by the descriptions of being in fog on the Grand Banks detailed in the subsequent section of this chapter – are not simply imaginative. While humans do not have dedicated physical mechanisms to detect changes in the moisture and humidity of an environment similar to those of insects (for which these are an important environmental sensing mechanism), Filingeri (2015, 763) has identified that in humans "the sensory integration of cutaneous thermal (i.e., evaporative cooling) and tactile (i.e., mechanical pressure and friction) sensory inputs" function to detect skin wetness and humidity and, thereby, perform a similar hygroreceptive function. Feelings of clamminess are therefore very real and, in some individuals, can lead to senses of claustrophobia, of being "hemmed in" by humidity. And then there's the auditory element. The water particulates that constitute fog impair human auditory perception, inhibiting the transmission of sound and making sounds fainter and, thus, less easy to perceive, and also blurs their timbres and attack, making their nature and sources both ambiguous and ill defined. Fogs are also *tricky*. As Shagapova and Sarapulova (2014, 62) have identified, there is no necessary correlation between the density of fog and the attenuation of sound, "when the mass content of water in drops increases within a certain range, a damping coefficient may decrease; i.e., a thicker fog may turn out to be more acoustically 'clear'" – a phenomenon that is confusing in itself. Similarly, "random inhomogeneities" within fogs may contribute to the apparent "scattering" of sound in foggy spaces (Rozenfeld 1983, 231). Combined with the disorientations produced by visual impairment and related perceptual-cognitive difficulties, the simple experience of being *in* fog can be perceived as fraught even before issues of the safe management of the situation and/or tasks ascribed during it are considered.

Dense fog's ability to cloak objects close to ships, combined with its aforementioned capacities to distort sound, and an intensified human concentration on sound in such vision-impaired environments, have lead to particularly disconcerting experiences on the Grand Banks in the Spring-Summer period when drifting ice is at its peak. Aspects of this were calmly (under)stated in reports by US Coast Guard ice patrol vessels sent to the Grand Banks in 1920 to monitor ice incidence in the form of large icebergs and smaller fragments known as *growlers* (so-called due to the animalistic sound of trapped air that escapes from them as they melt) in shipping lanes. Lieutenant Commander John Boedeker, captaining the cutter *Androscoggin* in April–May, for instance, noted that after a brief period of clear weather on April 30th he steamed northward to 43° 44' N, 48° 55' W until:

> we passed a few small pieces of ice close alongside, and stopped at once to await daylight, as it was thought that this ice indicated the presence of a berg or a growler in the vicinity. When completely stopped and drifting, the swash and breaking of the sea against the ice could be distinctly heard about two points on the bow and very close aboard. The sound was unmistakable, but the fog was too dense to make out the berg. At 4.20 a.m., there was a loud, heavy rumbling like thunder, accompanied by a heavy bumping noise and splashing [as a section of ice broke off and fell into the water].
>
> *(US Coast Guard 1920, 11)*

Fortunately, the *Androscoggin* was unaffected by this "close call" but the sonic scenario enacted in the dark, and the material presence it indicated as looming close by in the fog, is a vividly dramatic one.

As previously outlined, the drive for great productivity on the Grand Banks led to increased numbers of small dories being launched from schooners on the banks, usually crewed by two fishers. The boats carried lines, tools, bait, food and water for the crew and foghorns and/or other forms of noisemakers (including conch shells[6]). The latter were necessary in case the dory lost sight of its mothership in the fog and had to signal its whereabouts in order for the mothership to try to locate and retrieve it. Crew remaining on the schooners were tasked with listening out for dory noises and producing their own, using similar equipment, to help the dory crews to locate the ships and try to approach them. However, as the previous discussion of sound transmission in fog outlined, this was not a straightforward procedure – especially in stormy conditions that were both noisy in themselves (with thunder, wind, crashing waves) – and that made it difficult for dories and ships to locate and approach each other. Additional difficulties arose in fog when other vessels were sailing through fishing areas, with the risk of collisions that were commonly fatal for dorymen. There are extensive examples of dory crews that could not reconnect with their motherships by nightfall and that had to wait until morning to try to use the increased visual opportunities delivered by daylight to find their ships. Others spent longer periods adrift before being picked up by their parent ship or another vessel (which could result in them being carried back to a remote port and having to find their way back home independently), and there are many examples of crews who simply disappeared, presumably drifting out of populated fishing areas and shipping lanes. Even for those crewmembers who returned to their motherships, exposure, frostbite and/or injuries were common.

While the early phases of the fishing revolution enacted on the Grand Banks were largely an anonymous, unheralded exercise, the experiences of hardy fishers out on the stormy, fogbound seas were increasingly represented in various media in the late 1800s (in a period when dory fishing was already in decline). These experiences were ably represented by the New England painter Winslow Homer,[7] who produced a number of significant maritime-themed works after sailing on schooners operating out of Gloucester and talking with the crews about their experiences (Provost 1990, 20). One of his best-known works is *The Fog Warning* (1885), which shows a schooner on the horizon dwarfed by an ominous fog bank rolling in towards it and a solitary fisherman in a dory gazing back at both. The fisherman's concerned glance suggests an anxiety about whether he will reach his mothership before it becomes obscured from view. Complementing this painting, Homer's work *Lost on the Grand Banks* (Fig. 4.2) (also 1885) ably illustrates

FIGURE 4.2   Winslow Homer's *Lost on the Grand Banks*. (Wikimedia Commons, 1885)

key themes explored in this chapter by representing a dory pitching on a wave with no mothership in sight. Its crewmembers are depicted staring intently off-canvas, as if straining to see their schooner in the foggy gloom, with the colours of water and sky rendered in similar palettes and merging at the mid-right of the image. While these two paintings are not direct visual records of actual events, they are substantially researched and are thereby insightful about issues regarding the precarity of fishing on the Banks, and they convey the theme of stoicism in the face of adversity that attracted artists to represent them.

Along with Winslow's paintings, perhaps the best-known representation of fishing on the Grand Banks was supplied by British journalist and novelist Rudyard Kipling, who resided in Brattleboro, Vermont, between 1892 and 1896. Intrigued by tales of the New England fishery related by a local doctor named James Conland, who had worked on it in the 1850s–1860s, Kipling conducted research in Gloucester and Boston, familiarised himself with charts of and general information about the Grand Banks (Ormond 1995, xii–xvii) and dramatised aspects of these in his novel *Captains Courageous*, published in 1897. One of the most striking aspects of the novel, for this chapter at least, is the representation of fog, particularly in Chapter 5, which sets the scene of fishing on the Grand Banks, which he describes as "a triangle two hundred and fifty miles on each side a waste of wallowing sea, cloaked with dank fog, vexed with gales, harried with drifting ice, scored by the tracks of the reckless liners, and dotted with the sails of the fishing-fleet".[8]

Given that the volume of schooners fishing on the Banks had declined by the 1890s, the crowded scene is perhaps more representative of an earlier period, but the description of the "dank" cloaking fog is astute and his extended description of its effects has some credibility:

> For days they worked in fog—Harvey at the bell—till, grown familiar with the thick airs, he went out with Tom Platt, his heart rather in his mouth. But the fog would not lift, and the fish were biting, and no one can stay helplessly afraid for six hours at a time. Harvey devoted himself to his lines and the gaff or gob-stick as Tom Platt called for them; and they rowed back to the schooner guided by the bell and Tom's instinct; Manuel's conch sounding thin and faint beside them. But it was an unearthly experience, and, for the first time in a month, Harvey dreamed of the shifting, smoking floors of water round the dory, the lines that strayed away into nothing, and the air above that melted on the sea below ten feet from his straining eyes. ... They made another berth through the fog, and that time the hair of Harvey's head stood up when he went out in Manuel's dory. A whiteness moved in the whiteness of the fog with a breath like the breath of the grave, and there was a roaring, a plunging,

and spouting. It was his first introduction to the dread summer berg of the Banks, and he cowered in the bottom of the boat while Manuel laughed.

Homer's painting and Kipling's novel merit reference in this chapter given that the everyday experiences of fishermen from isolated coastal areas were rarely – if ever – regarded as significant (and/or directly reported) outside of the oral culture of their home communities or the broader fishing community in the period. Those accounts that have survived from this context primarily take the form of songs and verses that were collected by early- to mid-20th-century folklorists. Elizabeth Bristol Greenleaf and Grace Yarrow Mansfield, for instance, who collected in northern Newfoundland in the 1920s, heard "many songs… about life while out fishing on the Banks" (Greenleaf & Yarrow Mansfield 1933, 228), including ones that celebrated the adventure of the experience (such as 'The Banks of Newfoundland') and others that recorded fishers separated from their vessels and/or "the loss of ships and men" (such as 'The fishermen of Newfoundland/The good ship Jubilee') (Greenleaf & Yarrow Mansfield 1933, 285–287). Historians, folklorists and local writers who worked in areas of regional Newfoundland in the mid-20th century were also given various accounts of dorymen lost for days, weeks and even months before being able to return to their home ports by either the protagonists or members of their communities. The story of George May and Charles Williams, who sailed to the Grand Banks in 1927 on a schooner operating out of Fortune Bay, as recounted in Newfoundland historian L.W. James's edited 1961 collection *The Treasury of Newfoundland Stories*,[9] provides one such example:

> They had not gone more than a few hundred yards from the vessel when the fog, with the suddenness for which it is noted on the banks, closed down upon them. Bank fishermen are well-accustomed to thick fog, however. May and Williams continued on their way toward the trawls. They were unable to find them, though they spent an hour searching where they thought them to be. They decided to return to the ship, which was to prove as elusive as the trawls. For five hours they rowed and sailed, but never a sign or sound of her greeted them. The dreaded realization dawned in their minds that they were, for the time being at least, quite hopelessly lost.
>
> *(Unattributed 1961a)*

Realizing that they were unlikely to rendezvous with their ship, they attempted to navigate their way back to St. John's (Newfoundland) with the aid of a small compass. Either their navigation was awry or they were being carried away by currents, but by day five – with the heavy fog still in place – they were tired, severely afflicted by the cold and wet and suffering from thirst. Fortuitously they came upon an iceberg and managed to break off ice to suck. Weak and hungry, they were eventually discovered on day 11 by the steamship *Albuera*, 600 kilometres from where they had initially become lost, and they were given passage to the English port of Tilbury (Unattributed 1961a). Other dorymen were not so lucky, such as William Butt, the companion of William Strickland, who set out to the Grand Banks from Burin on a schooner in April 1897. Once on the Banks their dory departed from its mothership on a "fine but misty morning" before the weather turned and they found themselves lost in an increasingly windy "thicken-fog" (Unattributed, 1961a). A not uncommon fate awaited them.

> When darkness came the wind was a gale from the southeast and the sea was running high; it required all our skill and strength to keep our boat from filling. At about 10 o'clock we heard a steamer's fog hooter but, although we pulled frantically in the direction of the repeated sound, which seemed for awhile to be looming nearer to us, no lights of a receiving steamer were seen and the mournful bellowing died away in the dark distance. Again, perhaps near midnight, another whistle – clear and distinct – was heard and our hopes were raised only to be shattered once more.
>
> *(Unattributed 1961b)*

The next six days proved traumatic. After sighting land low on the horizon on the fourth, the pair rowed towards it, only for Butt to expire. After two more days afloat, and severely weakened, Strickland was finally rescued by a fisherman and carried ashore at Ramea Island, off the southcoast of Newfoundland.

Along with these stark accounts of misadventure, one of the most detailed subjective descriptions of dory fishing in adverse conditions was provided by Australian sailor and journalist-adventurer Alan Villiers in 1952 when he tried (sole) crewing a dory from a Portuguese ship named the *Argus* "for six cold and foggy weeks" in Spring 1950.[10] I quote at length to convey the experience Villiers recalls having in the fog whilst trying to return to his mothership:

> that infernal fog had blown down. An arm of it was between me and the Argus. Suddenly I found myself alone on the sea, and the ghostly arms of the horrible fog were wraithlike around me ...
>
> After a while I heard a schooner sounding her great fog siren, for they all carried air-raid sirens at their mastheads to summon the dorymen in fog. But where was this schooner? Was she the Argus? I didn't know. I had no judgment of sound direction in fog.

I knew the distinctive signal the Argus used on her fog bell, the big old church bell which hung in the mizzen rigging. I blew this signal on my conch shell and waited for a reply.

None came. Then that siren wasn't aboard the Argus, or I'd heard it through some freak condition in the fog. I carried on. I blew my conch again.

What was that? An echo? I blew again. It was no echo! It was a doryman sounding our signal, an Argus man, in the fog like myself. I shouted to catch his attention, to check my compass course with his. He shouted back. I saw nothing but the white and ghastly fog and the greasy cold swirling of the wretched sea.

And then, indistinct at first and almost unbelievable, I saw the triangle of a tiny sail harden in the surrounding murk; a roll of white water gurgled at the little laden bow. There was a dory! No. 16!. …Once I had found the First Fisher, I knew I was all right. After a quarter of an hour of gliding through the pall of fog, a white monstrosity suddenly loomed above us, right alongside. It looked like an iceberg …

"Argus," Battista grinned. "We come back, no?"

(Villiers 1952, 596)

Villiers' account suggests that his hygroreceptivity led him to feel uncomfortable, disorientated in and even menaced by the fog (as conveyed in phrases such as "the ghostly arms of the horrible fog were wraithlike around me" and "the white and ghastly fog") compounded by the anxiety about losing the mothership, its safety and comforts. While colourfully phrased, the senses of stress he records are all the more credible since he was an experienced sailor in cold water climes waters, having previously sailed on Antarctic whaling ships in the Ross Sea and having written of these experiences in his book *Whaling in the Frozen South* (1931) in a relatively prosaic manner.

Although it peaked in the mid- to late 1800s, dory fishing from schooners continued until the 1960s, when it was finally superseded by modern ships that were equipped with sonar, radar and electronic navigation systems that enabled them to more precisely target fish masses from single vessels. Large, industrial-scale fleets of ships briefly dominated the fishery before they decimated regional fish stocks, leading the Canadian Government to declare a moratorium on fishing in the area of the Grand Banks that fell within the 370 kilometre (200 nautical mile) Economic Expansion Zone it had declared in 1992. The moratorium has been maintained, albeit with variation and no little controversy, to the present (Rose & Rowe 2022).[11] The period of the dory fishery discussed in this chapter was thereby just over a century in duration and is now remembered, with some nostalgia, as a time when fish were plentiful and when human grit and labour could secure favourable returns. The sensory confusion, stress and frequent injury and loss of life experienced by dorymen is less often recalled but is nonetheless significant as an example of the manner in which capitalist enterprises can require and inveigle individuals into working in extreme conditions – in this case, into one of the foggiest, stormiest (and frequently sea-ice populated) areas of the planet's oceans. In this regard, we might view the dorymen's experiences as an effective experiment into human performance under stress and disorientation whose results were recorded, subjectively and haphazardly, in a range of cultural media that can provide only partial glimpses of and speculations about experience. This history is, nevertheless, significant in itself – as a cluster of vivid experiences – and as another example of extractivist capitalism's demanding deployment of labour in harsh and demanding environments while over-exploiting the resource concerned.

Fog, Flight and Oil Rigs

While the fog, storms, ice and all-encompassing cold may no longer play havoc with dorymen's lives, the decline of the Grand Banks fishery was followed by a new form of extractivist exploitation in the eastern part of the area, the development of oil drilling, led by the massive Hibernia platform, which commenced production in 1997 after a fraught exploration and construction process in which the area's full range of environmental hazards had to be negotiated. One of the most notable incidents was the loss of the Ocean Ranger in 1982, a semi-submersible drilling unit that sank with 84 crewmembers onboard, leaving no survivors. Subsequent wells have been developed nearby on the Terra Nova and White Rose fields and local conditions have continued to cause considerable logistical difficulties for their operations, particularly in terms of rotating staff and provisioning by means of ships or helicopters. Difficulties in landing the latter on platforms during foggy daytimes have led to frequent cancellations or turn-backs (and operator pressure to compensate with – equally problematic – flights on clear or semi-foggy nights) and other weather problems have also contributed to a number of crashes by helicopters working the route. These issues were explored in the 2010 Canada-Newfoundland and Labrador Offshore Petroleum Board (C-NLOPB) inquiry into the 2009 crash of a Cougar Helicopters flight returning from a platform that resulted in the death of 17 of those on board. Along with its examination of procedural and engine part issues, the C-NLOPB report documents that piloting and travelling on helicopters over the Grand Banks to rigs located on it is a singularly risky exercise. The difficulty of predicting and safely negotiating fog patterns around the Grand Banks oilfields has been documented and analysed by Bullock, Isaac, Beale and Hauser (2016) and Bodaghkhani (2017). Referring to his case study area – the Hibernia field – in particular, Bodaghkhani (2017, 20) has contended that fog:

> forms and develops due to multiple local microphysical, dynamic, and radiative processes; these are in turn influenced by boundary layer and synoptic-scale meteorological conditions. ... The ways these various influences work in combination and opposition varies considerably between locations, and consequently understanding fog and improving predictability typically requires detailed research on specific locations of interest.

Until such detailed observations, predictive models and monitoring of fog for incoming and outgoing air transport are developed, the precarity of helicopter transit between Grand Banks rigs and shores can be considered alongside dorymen's experiences on the open seas as a direct effect of working in such fraught environments.

Conclusion

As this chapter has detailed, the Grand Banks are as singularly inhospitable to human perceptions and activity as they are "extraordinarily hazardous" to the conduct of marine livelihood enterprises in general. The creation of such a workplace in the stormy, fogbound area of ocean has resulted from high commodity prices and product demands, initially from the European market for fish and, more latterly, from the global thirst for oil in the final phase of a carbon dependent global economy. With particular regard to the working conditions of dorymen in the aquapelago, which have been the principal focus of this chapter, the experience of being upon the seas in fog and, more broadly, of inhabiting the affective atmosphere of a foggy oceanic space, represents a unique episode in human history. Capitalism can be regarded as having provided the stage, script and directions for a group of human actors to interact with the complex actant system of the northwest Atlantic. Neither the dorymen's experiences – let alone multiple losses of life, injury and/or psychological trauma – are *incidental* aspects of the fishery; they are key aspects of the human experience of the industrial operation in the aquapelago and the low value industrial enterprises place on human safety and comfort. The role that fog played in this drama is notable. While seemingly a softer, less aggressive agent than storms, raging seas or icebergs, its disruption of human perceptions and abilities to navigate space reveal it to be singularly problematic for humans engaging with it for extended durations. The study of affective atmospheres and human perceptions, disorientation and stress in an aquapelagic space detailed in this chapter attempts to centre the former with regard to livelihood activities as the *raison d'etre* of such immersions and to keep the focus on the extractivist enterprise more generally as it bends human actors to its will.

Acknowledgements

Thanks to Peter Narváez for inviting me to Newfoundland and introducing me to various aspects of the cod fishery in the 1990s and also to Jim Payne for illuminating conversations about maritime life in the region.

Notes

1 Fog – at sea and on land – is commonly associated with calm conditions, but it can often aggregate in windy conditions if temperature variations and air humidity are sufficiently marked.
2 Precisely these kind of conditions caused the "perfect storm" represented in Sebastian Junger's (1997) eponymous account of the loss of the Gloucester fishing vessel the *Andrea Gail* with all hands in 1991 (subsequently adapted into an eponymous film, directed by Wolfgang Petersen, in 2000).
3 Indeed, the West Indies were linked by an exchange of salt cod, intended for enslaved Africans working on plantations, and molasses, which were subsequently exported to Newfoundland and used for both food and rum production (Tye 2015).
4 See, for example, Myers, Hutchings and Barrowman (1997) or Kurlansky (1997) for a more generalistic overview.
5 While there is little scientific evidence (or belief) that aquatic biota on the Banks can return to its former levels in any foreseeable future, it is not impossible.
6 Conch shells were used to signal for various purposes in Newfoundland, including when small fish arrived inshore, hence their common description as 'bait horns'.
7 Homer was a New England landscape painter who turned to marine themes after visiting the United Kingdom in 1881–1882 and painting fishing themes around Cullercoats, a coastal town on Tyneside associated with the North Sea fishery and home to a small artists' colony (Newton 2001). Inspired by the topic, he returned to the United States and based himself in Prouts Neck, Maine, for the remainder of the decade, working on various fishery-themed canvases.
8 This quotation is derived from the unpaginated Chapter 5 of the online version detailed in the references section of this chapter.
9 The authorship of individual entries in the collection is not specified.
10 One of the most significant audiovisual representations of dory fishing on the Grand Banks, John O'Brien's CBC documentary *The white fleet: Portuguese fishermen on the Grand Banks of Newfoundland,* dates from 1967. The dories depicted in the film, many of the items carried in them and the fishing and processing practices are closely similar to Villiers's account and also recall Homer's (previously discussed) paintings.
11 Also see Thornhill Verma (2019) for a historical overview of the determinations and implications of over-fishing in the region.

References

Bodaghkhani, Elnaz. 2017. *Climatological perspectives on fog from the Hibernia Platform*. Master of Science thesis, Memorial University of Newfoundland.

Bristol Greenleaf, E. & Yarrow Mansfield, G. 1933. *Ballads and sea songs of Newfoundland*. Hatbro, Pensylvannia.

Bullock, Terry, Isaac, George A., Beale, Jennifer & Hauser, Tristan. 2016. *Improvement of visibility and severe sea state forecasting on the Grand Banks of Newfoundland and Labrador*. Paper presented at the Arctic Technology Conference, St. John's, Newfoundland and Labrador, Canada, October 2016. Paper Number: OTC-27406-MS.

Canada-Newfoundland and Labrador Offshore Petroleum Board. 2010. Offshore Helicopter Safety Inquiry. http://oshsi.nl.ca/userfiles/files/HELJA12.pdf

Filingeri, Davide. 2015. Humidity sensation, cockroaches, worms, and humans: Are common sensory mechanisms for hygrosensation shared across species? *Journal of Neurophysiology 114*(2), 763–767.

Gilbert, William. 2011. Beothuk-European contact in the 16th century: A re-evaluation of the documentary evidence. *Acadiensis 40*, 24–44.

Griffero, Tonino. 2019. Is there such a thing as an "atmospheric turn"? Instead of an introduction. In *Atmosphere and Aesthetics: A Plural Perspective*, edited by Tonino Griffero and Marco Tedeschini. Palgrave Macmillan, 11–62.

Guerin, Ayasha. 2019. Underground and at sea: Oysters and Black marine entanglements in New York's Zone-A. *Shima 13*(2), 31–55.

Harley, Erin M., Dillon, Allyss M. & Loftus, Geoffrey, R. 2004. Why is it difficult to see in the fog? How stimulus contrast affects visual perception and visual memory. *Psychonomic Bulletin & Review 11*, 197–231.

Hodges, Benjamin Kidder. 2022. Atmospheric visions: Mirages, methane seeps and 'clam-monsters' in the Yellow Sea. *Shima 16*(1), 115–136.

Holm, Paul, Ludlow, Francis, Scherer, Cordula, et al. 2018. The North Atlantic fish revolution (ca. AD 1500). *Quaternary Research 108*, 1–15.

Junger, Sebastian. 1997. *The perfect storm*. Norton.

Kipling, Rudyard. 1897. *Captains courageous*. Doubleday. http://www.telelib.com/authors/K/KiplingRudyard/prose/CaptainsCourageous/captcourage

Kurlansky, Mark. 1997. *Cod: A biography of the fish that changed the world*. Vintage.

Martin, Craig, 2011. Fog-bound: Aerial space and the elemental entanglements of body-with-world. *Environment and Planning D: Society and Space 29*, 454–468.

Maxwell, Ian. 2012. Seas as places: Towards a maritime chorography. *Shima 12*(1), 22–24.

Myers, Ransom A., Hutchings, Jeffrey A. and Barrowman, Nicholas J. 1997. Why do fish stocks collapse? The example of cod in Atlantic Canada. *Ecological Applications 7*(1), 91–106.

Newton, Laura. 2001. *The Cullercoats artists' colony c1870–1914*. PhD thesis, University of Northumbria at Newcastle.

O'Brien, J. 1967. *The white fleet: Portuguese fishermen on the Grand Banks of Newfoundland*. CBC NL. https://www.youtube.com/watch?v=ELfR82zk08Y

Ormond, Leonee. 1995. *Introduction to Rudyard Kipling: Captains courageous*. Oxford World Classics, xiii–xxxx.

Provost, Paul Raymond. 1990. Winslow Homer's the fog warning: Fisherman as heroic character. *The American Art Journal 22*(1), 20–27.

Rankin, K.J. and Holm, Poul. 2019. Cartographical perspectives on the evolution of fisheries in Newfoundland's Grand Banks area and adjacent North Atlantic waters in the sixteenth and seventeenth centuries. *Terrae Incognitae 51*(3), 190–218.

Rose, George A. and Rowe, Sherylynne. 2022. Congruence of stock production and assessment areas? An historical perspective on Canada's iconic Northern cod. *Canadian Journal of Fisheries and Aquatic Sciences*. https://cdnsciencepub.com /doi/10.1139/cjfas-2021-023

Rozenfeld, S.K. 1983. Scattering of sound waves by random inhomogeneities of atmospheric fog. *Akusticheskii Zhurnal 29*, 392–397.

Serres, Michael. 1985. *The Five Senses: A Philosophy of Mingled Bodies.* (2008), Bloomsbury.

Shagapova, V.S. and Sarapulova, V.V. 2014. Features of sound refraction in the atmosphere in fog. *Izvestiya, Atmospheric and Oceanic Physics 50*(6), 602–609.

Stoodley, Allan. 2021, November 14. How Grand Bank made the 'Grandy dory' a Newfoundland icon. CBC Newfoundland and Labrador. https://www.cbc.ca /news/canada/newfoundland-labrador/grandy-dory-newfoundland-1.6225449

Thornhill Verma. Jennifer. 2019. *Cod collapse: The rise and fall of Newfoundland's saltwater cowboys.* Nimbus Publishing.

Tye, Diane. 2015. "A poor man's meal": Molasses in Atlantic Canada. *Food, Culture & Society 11*(3), 335–353.

Unattributed. 1961a. Adrift in an open dory. In *The Treasury of Newfoundland Stories,* edited by Lemuel Willey James. Maple Leaf Mills Limited. http://ngb .chebucto.org/Articles/adrift.shtml

Unattributed. 1961b. Adrift on the Banks in an open dory six nights. In *The Treasury of Newfoundland Stories,* edited by Lemuel Willey James. Maple Leaf Mills Limited. http://ngb.chebucto.org/Articles/adrift2.shtml

United States Coast Guard. 1920. *International ice observation and ice patrol service in the North Atlantic Ocean: Season of 1920.* Treasury Department.

U.S. Department of Homeland Security Navigation Center of Excellence. n.d. Weather. uscg.gov/?pageName=IIPWeather#:~:text=Foggy%20conditions%2 0can%20exist%20most,foggy%20conditions%20with%20calm%20winds

Vale, Celina. 2018. Understanding island spatiality through co-visibility: The construction of islands as legible territories - a case study of the Azores. *Shima 12*(1), 79–98.

Villiers Alan J. 1931. *Whaling in the frozen South: Being the story of the 1923-24 Norwegian whaling expedition to the Antarctic.* Robert McBride.

Villiers, Alan, J. 1952, May. A pesca do Bacalhau: I sailed with Portugal's captains courageous. *National Geographic,* 565–596.

Wikicommons. 1885. Winslow Homer's 'Lost on the Grand Banks'. https://en .wikipedia.org/wiki/Lost_on_the_Grand_Banks#/media/File:Lost_on_the_Gra nd_Banks_by_Winslow_Homer_1885.jpg

Wikicommons. 2006. Flemish cap (map). https://en.wikipedia.org/wiki/Flemish _Cap#/media/File:Grand_Banks.png

5

COLONIAL LEGACIES AND RESTORATION FUTURES

Examining the Risks of Dispossession from Coral Reef Restoration in the Indonesian Aquapelago

Jessica Vandenberg

Introduction

In late 2016, a small island community in the Spermonde Archipelago of Indonesia learned that they had been selected by a multinational corporation, in conjunction with regional authorities, as the site of a large-scale coral reef restoration (CCR) initiative. Several islands self-nominated and were surveyed by regional university scientists to determine the most socially supportive and ecologically suitable site for CRR. Based on this assessment, the aforementioned island community was selected. An official ceremony, attended by invited government officials and company representatives, was held on the island to grant the community their newly designated title, and also to explain how over the next several years their surrounding coral reefs would be transformed. The reefs would be restored, the fish would return and the community would reap the benefits. Since the deployment of the restoration program, coral has begun to grow back, and fish are slowly returning; however, these ecological changes have not necessarily translated into the promised social benefits. Instead, some members of the local community perceive they have been dispossessed of their rights to the surrounding seascapes, while others are fearful of future dispossession from their island home.

This particular narrative, where local people are displaced and dispossessed of resources, landscapes and seascapes on which they historically depended, is a familiar one in the conservation world. Efforts to sustainably manage vulnerable biodiversity have increased substantially in recent decades due to the recognition of anthropogenic change in the biosphere. Yet major social costs have accompanied these initiatives. Dispossession, similar

DOI: 10.4324/9781003569534-5

to what is described above, has been argued to be a recurring theme of conservation globally (West 2006). For example, the establishment of protected areas for the purpose of protecting vulnerable species or ecosystems has led to the displacement of tens of millions of people from the landscapes and seascapes where they have historically resided, farmed, hunted, fished and foraged (Agrawal & Redford 2009). While some protected area programs have led to successful social and economic outcomes (Persha, 2011), many scholars argue that this process prioritizes the conservation of rare species and/or vulnerable ecosystems over social equity and human welfare; and is thus a new form of "accumulation by dispossession" (West 2016; Corson & MacDonald 2012). This unintended consequence is, however, paradoxical as many forms of conservation aim to simultaneously achieve both conservation and development goals (McShane & Wells 2004). Despite heavy criticism of these social consequences and sustained goals to improve the wellbeing of local people, protected area approaches to conservation remain a mainstay globally and, in some cases, continue to lead to various forms of dispossession.

More recently, studies on dispossession and conservation have extended beyond the terrestrial to aquatic spaces and dispossession has been widely referred to as "ocean-grabbing", a process defined as "the dispossession or appropriation of use, control or access to ocean space or resources from prior resource users, rights holders or inhabitants" (Bennett et al. 2015, 62). Studies on ocean-grabbing and its social implications have focused on livelihood and human security impacts (Barbesgaard 2018). Affected communities can be dispossessed of not only the rights to space and marine resources but also to other forms of non-material rights that are vital to small island and coastal life. Illustrated through a case study of corporate-led CRR in the Spermonde archipelago of Indonesia, this chapter describes how this particular marine conservation program led to various fears about dispossession, including but not limited to marine spatial access.

CRR is the process of assisting coral ecosystem recovery from a state of disturbance to a state where their structure and function is self-sustaining (Edwards 2010). It is often presented as a marine conservation solution that provides "win-win" outcomes, where biodiversity and food security objectives can simultaneously be realized (Hein et al. 2019). In some instances, successful outcomes occur in both domains (Kittinger et al. 2016); however, like other approaches to marine conservation, CRR can create unintended negative consequences for the communities designed to benefit from it. Similar to marine protected area (MPA) approaches to marine conservation, coral restoration programs adopt no-take fishing restrictions and coral protection measures, to protect and maintain reef restoration infrastructure and coral transplants. Establishment of no-take marine protected areas has been found in some cases to lead to the dispossession of local peoples' rights to their surrounding marine resources, impacting their livelihoods and food security (Darling 2014). This often results when MPAs are not well supported by the local community (Bennett & Dearden 2014). West (2016) explores this process of dispossession further by describing the ways that conservationists tend to delineate local communities' resource governance practices and environments as "prior nature" and "prior practices". She notes how this articulation facilitates dispossession by devaluing local knowledge and practices, producing and reinforcing inequality.

Utilizing West's framework of dispossession to emphasize how modes of engagement create both material and non-material forms of dispossession, I explore three main processes of dispossession: the loss of rights to aquapelagic (i.e. integrated terrestrial and marine) territory (Hayward 2012), the further marginalization of local people and their social networks (that result from dispossessive processes) and the deterioration of community security and wellbeing (Lowe 2013; West 2006, 2016).

I approach this topic through an aquapelagic framework because the interconnectedness of small island life is foundational to this region of Indonesia and because entanglements between islands, island people and oceans are a critical factor of dispossession in many marine spaces of the world. An aquapelagic framework is well suited to assess ocean-based dispossession because a fundamental part of thinking aquapelagically is the notion that an aquapelago is "an entity constituted by human presence in and utilization of the environment (rather than as an 'objective' geographical entity)" (Hayward 2012, 6). Dispossession does not merely occur through physical displacement of local people from their homes or from preventing access to marine space. It can also take shape through prohibiting socially meaningful human-environment interactions, such as fishing and navigation, and inhibiting reciprocal social relations that are centerpieces of small island life. Therefore, thinking aquapelagically about these complex entanglements allows for greater insight into the potential consequences of CRR, a conservation practice that is only increasing in popularity. It should also be noted that the aim of this chapter is not to denigrate restoration or to criticize this particular project as intentionally malicious or having coercive intent. The company implementing this particular project aims to provide social benefits to the community that are locally valued. It has attempted to adapt its engagement to be more participatory to rectify the unintended dispossessive results of the program, signifying their willingness to improve engagement practices and address the challenges that come with working within complex human environmental systems. This chapter seeks to demonstrate how socially complex conservation development projects can be and how important concerns and fears can develop within the community. I illustrate how history, culture and power relations complicate seemingly simple ecological and social outcomes in order to urge conservationists and CRR practitioners

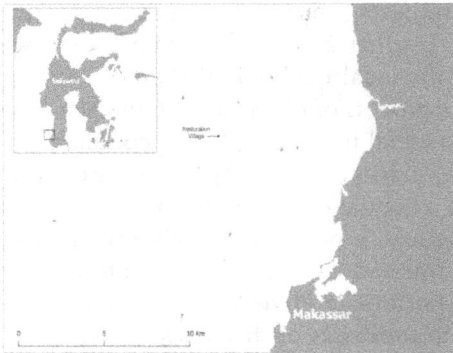

FIGURE 5.1   Map of the Spermonde Archipelago in South Sulawesi, Indonesia, showing the location of the Restoration Village and the "mainland" city of Makassar.

to re-think their own assumptions about small island communities and to consider these factors when developing reef conservation strategies.

This chapter draws from ethnographic field data I collected between 2016 and 2019, on a small island in the Spermonde Archipelago of Indonesia where this corporate-led CRR took place, hereafter referred to as the "Restoration Village" and three neighboring islands (Fig. 5.1).[1] The Restoration Village community has a population of approximately 1,100 people, all of whom are Makassarese, one of the four major ethnic groups that reside in this region. I observed and collected accounts of the ways the restoration initiative influenced small island life across these islands and in the broader Spermondes. Over four, 1–3 month field seasons, between May 2017 and August 2019, I conducted 124 household surveys and 30 semi-structured interviews with islanders. Additional data and insights were collected through participant observation and informal interviews with island residents. All surveys and interviews were conducted in Makassarese with the assistance of a local translator and utilizing best practices in informed consent.

The Spermonde Archipelago of Indonesia is located in the center of the Coral Triangle, a region known to have the highest coral and fish diversity on earth (Sanciangco et al. 2013). The archipelago is composed of approximately 180 coral islands, 54 of which are densely populated, and it is located approximately 60 km off the coast of Makassar, the capital city of South Sulawesi (Fig. 5.1). People in the Spermonde rely upon fishing as the dominant livelihood with an estimated 6,500 fishing households in the region (Pet-Soede et al. 2001). Most fishers are employed through a patron-client fishery system, which is characterized as a hierarchical wage-labor systems where the patron or boss (known locally as *pungawwa*) provides gear, boats, access to markets and loans to their client fishers or crewmembers (*sawi*)

(Ferse et al. 2012). Patron-client fishers of the Restoration Village are mainly involved in the *pa'gai* fishery. The term *pa'gai* refers to the type of boat used (20–30-foot-long vessel) and is crewed by a group of 10–12 men. *Pa'gai* fishers use purse seine nets to target pelagic species, such as mackerel and squid. They also seasonally fish the surrounding reefs for coral squid and cuttlefish. A fraction of fishers in the community are semi-subsistent independent fishers, heavily reliant on the local reef for their livelihoods. Beyond livelihood and food security benefits, surrounding reefs are locally valued for storm protection of their island home and their boats. It is also an important space for children to play – swimming, fishing for small coral fish or foraging for shellfish on the reef flats.

The CRR program began in July 2017 and posed as a dual conservation and development initiative. It is designed and implemented by a large multinational corporation that uses resources from the region. The stated objectives of the program are to conserve and restore the biodiversity of local coral reefs, while simultaneously improving the food security of the local community through increased fishing yields. Although the program is executed through this private company, academic research partners from several local and international institutions are involved and have provided guidance on the design and monitoring of the restoration program. The Restoration Village was selected as the project site through a regional government-supported selection process and project engagement is targeted towards the Restoration Village community, while neighboring villages are not included in project decision-making processes. Initially the village head (*kepala desa*) was consulted and agreed to the restoration project. Afterwards, an agreement letter was distributed to each household. This letter was intended to inform community members of the project and to garner support. This approach was not fully effective. Numerous interviewees stated that they either never received a letter because they were away fishing or that they had received a letter but could not read its content. Beyond this initial communication, community engagement by the company in the Restoration Village has occurred through public meetings in the village square, decision-making meetings with formal and informal village leaders, informational pamphlets and gifts to the mosque. Decision-making meetings are held regularly with a select few who may or may not share information within the community, whereas community-wide meetings to share the progress of the restoration project occurred twice between 2017 and 2019. Leaflets, brochures, posters and an annual calendar were also distributed, providing information on research findings from scientific studies conducted by academic collaborators investigating the marine ecology and coastal geomorphology of the island.

The restoration project employs a community-based model where residents are paid to assemble CRR structures, providing a short-term economic stimulus in the community. Initially neighborhood heads (*rukun tentangga*) selected participants (typically 6–10 people) from their neighborhood. However, according to several community members, this method was later abandoned in order to be more inclusive, allowing anyone within the community to participate if interested. Community members tie coral fragments to hexagonal-shaped steel structures, termed "spiders" at the time I studied the project, which are then deployed by trained divers to designated restoration sites around the island. On average, deployment events occurred monthly during the dry season, employing around 36 local men and deploying 550 spiders in areas 1,000m^2 over a three-day period. A fixed budget was allocated for monthly deployment events. Individual compensation therefore varied depending on the number of individuals who participated. On average, compensation was comparable to a typical day's wage as a fisherman in the community ($5–$7 USD).

Multiple members of the community believed that the restoration infrastructure was initially intended to be deployed within the existing bounds of an MPA established through the Coral Reef Rehabilitation and Management Project (COREMAP 1998-2013). However, through the life of the project, build sites expanded beyond the boundaries of the MPA across the reef crests and surrounding rubble fields. COREMAP is Indonesia's largest MPA initiative and is funded by the Asian Development Bank and the World Bank. The company hired two coral guards to enforce no-take restrictions of the COREMAP-established marine protected area and legally mandated restrictions on destructive fishing gear (i.e. bomb and cyanide fishing gear); however, these coral guards have been described by some community members as over-enforcing (compared to their COREMAP predecessors), restricting the use of spear guns and access beyond the bounds of the MPA. While it is not yet clear what the extent of these restrictions may be, some island residents have said that they are fully restricted from fishing around the island while others have maintained that they are only restricted to fishing where restoration infrastructure has been installed (and this restricted space has continued to expand since the project began). Although enforcing restrictions on fishers from neighboring islands is not alien to the region (Glaser et al. 2010), fishers from neighboring islands have been fully prohibited from fishing on the island only since the introduction of the restoration project. None of these new restrictions are officially mandated by the local government, and company project officers state that they do not require that coral guards enforce these newly adopted restrictions; however, the enforcement of these restrictions began with the implementation of this project.

Dispossession and Exploitation in the Spermonde

European imperialism in Sulawesi (and Indonesia) spanned over 300 years, beginning in the 16th century when European spice traders established processing factories in Makassar, the capital of South Sulawesi. Towards the end of the 17th century the Dutch claimed hegemony over the South Sulawesi region through a series of wars, pushing out all other European powers and allowing for monopolization of the spice trade in the Maluku Islands, east of Sulawesi (Knaap & Sutherland 2004). At this time settlement in the Spermondes was forbidden by the Dutch who designated these islands for Dutch naval use (Knaap & Sutherland, 2004). However, some historical accounts state that some of the islands had been settled by Bajau communities (Reid 1999); and, by the early 18th century, Malay, Indian and Arabic traders had settled on some of these islands, using them as a trading outpost (Mattulada 1994). When Indonesia gained its independence in the late 1940s, following the end of World War II, communities fled the islands due to political instability (Ferse et al. 2014). However, accounts from the Restoration Village community described that during the independence movement, violence in mainland Sulawesi pushed individuals to flee and some found refuge on the small coral islands of the Spermonde. The use of the Spermonde as a temporary space of refuge fits into the dominant cultural mode of the Makassar region of South Sulawesi. In Makassarese culture, coral islands are perceived as "amphibious" entities, meaning they are neither terrestrial nor aquatic (Gibson 2005). They are something in between landscape and seascape – liminal spaces, fit for a time of transition, or a place to find refuge. However, these communities continue to persist and the Makassarese people of the Restoration Village have adapted to small island life and fishing-dependent livelihoods, despite originating from agrarian backgrounds in mainland Sulawesi over two generations ago.

The Spermonde were then influenced by a second wave of migration in the late 1960s during Suharto's "New Order", which was largely focused on economic development and the corporatization of government to achieve broad political order (Knaap & Sutherland 2004). This period was also characterized by extreme violence. Genocide of ethnic groups, justified as "Communist cleansing" swept across Indonesia (Tsing 1993). Again, according to interview accounts in the Restoration Village, people seeking both refuge and economic opportunities resettled the Spermonde. Here, economic development took shape through the commercialization of fishing. New development opportunities and fishing technologies attracted fishers to the islands to partake in new forms of wage-labor fishing.

Fishing was appropriated from a practice that was subsistence-based to one that was commodity-based. This transition stripped local people of their autonomy and their rights to resources. The species that were targeted by commercial fisheries were those that could be sold in the urban markets of Makassar, but they did not fit into the local morality of production (i.e. the production of goods for local consumption). Furthermore, this new form of fishing resulted in a less diverse fishery, the depletion of select commercial species, the degradation of coral (as destructive fishing gear was adopted to target those selected species) and a poverty trap, where islanders no longer fished for their own food but to support the demand of consumers on the mainland and beyond (Gorris 2016). Under this wage-labor system "patrons" pay "client" fishers low wages but subsidize wages with loans. To pay off provided loans, fishers must sell their catch to patrons at below-market costs. This system makes it nearly impossible for fishers to ever regain their autonomy. Although there are evident costs to these systems, there are social benefits like financial and social stability in the community. *Pungawwa* thus serve important yet complex social and cultural roles and provide security for their *sawis* in time of hardship (Ferse et al. 2014).

Capitalist-driven resource appropriation extended beyond resource extraction to the design and implementation of biodiversity conservation across Indonesia. Most notably in the Spermonde, COREMAP, the large-scale MPA initiative mentioned previously, was designed and implemented as a program aimed at achieving both conservation and neoliberal development objectives. The development outcomes of COREMAP were designed to establish a "viable coral reef management system in Indonesia that places the community at the center of management" (Radjawali 2012, 547). In the Spermonde Islands, efforts were focused on establishing community-based initiatives, such as creating locally managed MPAs awarding grants to promote adoption of alternative livelihoods, such as aquaculture and tourism, and the development of coastal resilience infrastructure, such as breakwaters. Despite extensive efforts, many COREMAP-established MPAs remain unenforced, alternative livelihoods were poorly adopted and other management strategies were typically neglected once COREMAP representatives left host communities (Glaser et al. 2010). Furthermore, social inequities and community frictions developed because allocated project benefits were captured by the local elite. Ultimately, the greatest barriers to achieving COREMAP's goals stemmed from the lack of equitable collaboration and engagement with local communities and the inability to incorporate existing trade and social networks into conservation management strategies (Radjawali 2012). These particular challenges also highlight neocolonial legacies of conservation strategies in the region, where tools of reform are implemented through force and are expectant that local norms and values will change to Eurocentric ones. This history of colonial and neocolonial exploitation in the region that has shaped and reshaped the relationships between island, sea and people in the Spermonde lays the foundation for present-day interventions, such as the CRR program described in this chapter, and the ways they are received by local people in aquapelagic societies.

The following three sections discuss the multiple forms of dispossession that interviewees described having developed from the CRR initiative. These forms of dispossession are either direct experiences or related concerns about what is to come. The views that I describe are not necessarily held by the community as a whole, although they are more than minority views. Despite not being a consensual view across the community, they are critical to understand in order to achieve equitable social outcomes and to avoid the marginalization of groups through restoration practices.

Dispossession through Marine Appropriation

> Before the restoration project we saw the reef as "ours" and could fish freely; now we [ie some members of the community] do not feel as if we have ownership.
>
> *(Interview Respondent #5, July 2019)*

When discussing dispossession from land or marine territory, it is necessary to understand "who is being dispossessed of what and the types of rights and power they had to access property prior to dispossession" (Kenney-Lazar 2012, 1021). In the Restoration Village, some now perceive surrounding reefscapes once used for fishing and boat anchorages as having restricted access for local people, while coral restoration infrastructure is freely deployed. Local people believe that these reefs have effectively been appropriated for marine conservation, as their rights to access have been limited. Appropriation has been described as the centerpiece of the dual, related processes of accumulation and dispossession (Fairhead et al. 2012). It is the process where the ownership, use rights and control over resources that were once publicly or privately owned are transferred from the poor into the hands of the more powerful (Cernea 2006). This process perpetuates colonial and neocolonial legacies of "resource alienation" or "land-grabbing", and in this case, the appropriation of nature is made in the name of the environment, rather than in that of the State or economic development (Fairhead et al. 2012). In this way, ecosystems and their services adopt new forms of value and commodification that can be transactional in environmental markets, while local forms of value and dependence are ignored.

In the Restoration Village, the surrounding water and reefs have been appropriated as reef "stocks" that can be preserved and restored in an attempt to mitigate the strain that the company has put on the world's oceans and fisheries. Through the CRR and related sustainability initiative, the company has gained international recognition for their work in advancing corporate sustainability and now has clout in the space of marine conservation. In this context focusing attention on the restoration of reefs runs the risk of shifting the burden of guilt away from the company's own large-scale commercial fishing operations to the "destructive" fishing practices employed by local people.

Some local independent fishers perceive that they no longer have rights to access marine resources on most of the reefs that fringe their island, despite previously having and freely exercising that right prior to the restoration project. Although an MPA had been previously designated under COREMAP, this MPA was never strictly enforced. Furthermore, the new restrictions extend beyond the bounds of the COREMAP MPA to include most of the near shore waters of the island. Now, those fishers who once utilized the reef must travel farther, expending more fuel, and therefore more money, to catch the yields they had in the past. Some just fish in the limited space not included in the MPA and restoration areas and perceive that they catch less fish. Local spear fishers are no longer allowed access to any of the waters near the island and feel forced to consider new gear types. This new gear restriction is especially frustrating as these fishers observe divers from the CRR program on the reefs using spearguns to eradicate sergeant fish viewed as a pest for coral restoration. This particular consequence, where local fishers' livelihoods and potentially their food security have been adversely affected by restoration policy infrastructures, is somewhat ironic yet all too familiar in the conservation development literature (Walley 2010). The purpose of the CRR initiative was initially posed as a means of improving food security, where enhanced and healthy reefs would bring more fish for fishers to catch. However, excluding fishers from this area, a caveat that most respondents expressed was unbeknownst to them when they initially consented to the project, has led to reduced access to fishing grounds, reduced perceived fishing yields and reduced perceived food security for some families.

Finally, the question of spatial use and resource dispossession applies to more than fisheries. The removal of sand surrounding the island and on the island was banned by the community years ago as a method of mitigating the effects of sea level rise and coastal erosion. However, some local people perceive that the CRR project has free access to sand on the island. Sand is one of the materials used for the coral spiders, and the company views using the island's sand for CRR as acceptable, even though it is prohibited for building the homes in which local people live. The local people see how sand has been appropriated for conservation use, despite the importance of maintaining sand on the island to protect it from the threat of violent storms. They see conservationists prioritizing the production of healthy coral over their coastal resilience. One particular community member has been more outspoken over this issue than others. In an interview, two other respondents described how this community member was the only individual brave enough to protest this issue and call out the innate contradiction of allowing the conservation group to extract sand, while the community is strictly forbidden to do so. Although coral reef restoration is actually intended to help mitigate coastal erosion, the CRR project has not effectively communicated with the community that this is a potential ecosystem service and the project may also be damaging the community's own forms of mitigation by creating new methods without adequately consulting the community as to their intentions. Like restrictions on spear guns, sand exploitation restrictions are a clear reminder to the community of the double standard that exists between the company and the community, and that the conservation group has the power to decide which forms of resource use are appropriate and which are not.

Dispossession from Aquapelagic Social Networks

> There is a social problem between [the Restoration Village] and other islands because fishers are not allowed to use [a] speargun [here] anymore.
> *(Interview Respondent #5, July 2019)*

Reciprocal relations are a critical component of small island life in the Spermonde (Gorris 2016). They are a primary ingredient of the "marine adhesive" (Fleury 2013) that binds small island communities together across aquapelagic seascapes. Access to fishing grounds is often based on reciprocity, where providing access to marine territories controlled by one community can ensure access to territories managed by others. Social networks and alliances, such as reciprocal fishing relationships, are essential for ensuring food security and supporting livelihoods and wellbeing in rural communities that cannot exist in isolation (Fabinyi & Liu 2016). Interventions that interfere in these relations may lead to profound social consequences. The CRR project not only runs the risk of dispossessing fishers from the Restoration Village from rights to fish around their own island, but it also runs the risk of dispossessing local fishers of rights to participate in aquapelagic reciprocal relations, including rights to fish around neighboring islands. This under-explored form of dispossession resulted from the establishment and enforcement of the no-take restrictions in the reef restoration sites. Surrounding islands that are excluded from fishing on these sites protested these restrictions by prohibiting fishers from the Restoration Village from fishing in their surrounding marine territories. To repair these damaged social relationships between islands, four *pungawwa* forbade their *sawi* to participate in the restoration project to demonstrate a greater commitment to their inter-island relations, rather than the CRR initiative. These community-level acts of diplomacy were effective and, after much discussion and negotiation, many fishers regained access to neighboring island fishing grounds.

However, some community members are still prohibited from entering the waters surrounding other islands and are unwelcome at inter-island social gatherings. The islanders employed as coral guards are mainly the ones who experience these exclusions. These individuals were not trained in enforcement strategies and their efforts to keep potential fishers out of the island's waters have been described by others as aggressive and hostile and have sometimes led to near violent conflicts. In one instance, a group of fishers from another island, who were given direct permission from the head of the Restoration Village to fish, were forced to leave by coral guards. Later that week, a group of nearly 30 angry men, armed with spear guns, came to the Restoration Village and demanded to speak to the village head about the matter. The incident was frightening to many of the residents because near violent inter-island conflict is extremely abnormal. The wife of one of the coral guards involved in the incident expressed fear that her husband would be killed for his involvement in the project because of the negative view many of the residents of other islands had adopted towards him. His fishing gear has been vandalized on numerous occasions and he is still restricted from visiting some surrounding islands. This incident also indicates that some community members now view the CRR project as having greater authority over inter-island relationships than the village head. Coral guards view their responsibilities to the restoration project as superseding their responsibilities and relations with the local government and surrounding island communities.

So far, the CRR project has operated under the assumption that the program has limited isolated impacts on the community of the Restoration Village. They have ignored or are unaware of the "marine adhesive" that binds this community to others across the Spermonde. The appropriation of the Restoration Village's reefs for CRR continues to run the risk of dispossessing some fishers across this aquapelagic society of their rights to fish – not only in the immediate area of the project, but in other communities too – and it is at risk of continuing to dispossess them of the important reciprocal relations that are critical to small island life and that support livelihoods and wellbeing. It has also sparked fears about future forms of dispossession.

Dispossessive Infrastructure and Its Evocation of Fear

[The Company] has spent money and time on the project, maybe in the future when I am dead and gone, [the Company] will come back and own this island. They will return and claim the island from our children saying that their parents planted this coral on [the Company's] behalf and is now the property of [the Company] so they will have to leave. What [the Company] has built is a marker of ownership, which they can return to in the future and claim.

(Interview Respondent June 2018)

Some islanders have experienced lost autonomy over marine territory through their fear of the corporation, its power, and what coral restoration infrastructure represents. These community members mostly remained uninvolved in the project. From their perspective, damaging coral was the only explicit restriction enforced through the project; but given the power and influence of the corporation, they were fearful of the potential penalties. Heavy fines or imprisonment were a frequent concern for some boat captains upon returning to the island and dropping anchor and for fishers whose hooks and lures unintentionally snagged branching coral. They felt as if they had lost sovereignty over their island and its resources now living under the scrutiny of the corporation. This sentiment seems to be shaped by not only the new restrictions, but also by the coral restoration infrastructure itself. Many respondents expressed concern over the transactional relationship with the company where islanders were paid for preparing coral spiders. In hindsight, some worried that this exchange was in actuality payment for rights to the seafloor on which the spiders are mounted. They view the spiders as infrastructural markers of the company's territorial claim, a physical reminder of the company's presence and perceived ownership of the area.

These concerns are warranted and stem not only from a long history of colonial and neocolonial displacement and exploitation in the region, but also from news stories of displacement that flood their social media feeds. Intergenerational trauma of displacement, reinforced by news accounts of displacement, fuels some community members' concerns about the company's presence. As one community member put it, "it's impossible for [the company] to not expect anything in return after all that they have invested into the project". But it is the uncertainty of what the company expects in return that has left the community fearful for their future. Speculations range, reflecting what they have seen in other communities. The aquarium trade, rare minerals and tourism development are primary speculative drivers of the company's involvement. Fishers have described other islands where they no longer can fish because they are controlled by Westerners who have established aquaculture operations, while mining and tourism are classical narratives of past and present displacement in Indonesia.

The Spermonde is home to one of the largest aquarium fisheries in the world (Ferse et al. 2012). Some community members have expressed concern that the company is restoring the reefs surrounding the island to harvest coral for trade. The design of the CRR infrastructure itself is described by some respondents as a structure that might facilitate harvesting because coral pieces can easily be broken off of the rod iron frames. Tourism displacement theories were influenced by the regular flow of visitors from Western countries the community has frequently observed diving on their reefs and visiting their island since the restoration project began. Many community members perceive restoration efforts as a strategy to make the island attractive to wealthy tourists. Some view these potential tourism prospects as an economic opportunity to sell handicrafts or establish homestays, but others worry that increased tourism could lead to forced displacement of the community for development of a Western resort. This concern is reasonable given Indonesia's modern history of tourism-related displacement (Cole 2017). Academic researchers, nature filmmakers, science journalists, international politicians and corporate employees have been brought to dive the restoration site and bear witness its biological success. These visitors dive on the reefs, join local people in tying coral fragments onto spiders and "explore" the local village to experience "island life". Over the years the numbers of foreigners who have come to the island have increased, despite the company's claims that it is keeping visitation to a minimum, making the community concerned about what this growing interest might mean for their future.

Conclusion: Towards Historically Aware Aquapelagic Coral Reef Restoration

After three years studying this project, it appears that the seascape and island resources alike run the risk of being appropriated to be used, valued and experienced by the Westerners conducting the CRR initiative, not necessarily by the people in the community, despite claims to the contrary. Western researchers and other visitors are perceived by some community members to have the freedom to move about the surrounding reefs and water, while local people have been excluded from this space, even as the project designers state that they are restoring the reef for the future benefit of the islanders themselves. Despite its good intentions, the CRR initiative has led to social consequences that exemplify multiple iterations of "accumulation by dispossession" where the people who were expected to benefit from the supposed conservation as development promises are, in actuality, harmed by them (West 2016). Some local people perceive that they have lost autonomy over their reef resources to the multinational company. They see it in their inability to freely fish and navigate in their waters and in the ways that reef resources are managed and controlled by the company rather than their own local governance structures. They also experience it through their inability to freely engage and reciprocate in the aquapelagic society in which they are situated. This CRR approach has led to the initial dispossession of local people from space, rights and resources that are integral to small island life. Coral reef restoration and other spatially oriented marine conservation initiatives need to be more cognizant of the aquapelagic nature of small island societies and the potential ramifications that protected area management strategies and introduced marine infrastructures (material and policy-based) can have on small island communities. At the very least, CRR initiatives in archipelagic regions must adopt an aquapelagic framework that recognizes local and regional history, inter-island network systems and other social and cultural practices in order to move beyond the narrowly assumed benefits of coral restoration to equally assumed isolated island communities. Aquapelagically informed marine conservation strategies can expand how communities and their networks might be able to engage with private philanthropic CRR programs and their infrastructures in ways that do not drastically limit community autonomy and dispossess small island populations from the social and ecological relationships that sustain and protect them.

Acknowledgements

The author thanks first and foremost the communities of the Restoration Village for their willingness to engage in conversation and welcoming me into the community. I also thank Hamsani Hambali and Farhan Mutahar for their enormous support as translators, and their contributions to the development of this work. I also thank the UNHAS sociology department, especially Pak Ramli and Pak Suparman Abdullah, for their quality efforts to collect preliminary social data in the community. I also thank Ken Hamel for his assistance in creating the figures used in this chapter. Finally, I thank Amelia Moore and Carlos Garcia-Quijano for advising me through this work and providing comments on previous drafts of this chapter. All errors and omissions belong to the author alone.

Disclosure

The company involved in implementing the coral restoration program provided partial financial support for the research presented here. The manuscript of this chapter underwent an initial review period by the company to provide feedback on how the project is represented in this piece. This study's findings do not align with the company's views in all cases; however, a longer-term study, already underway, will be able to better address and contextualize these disagreements, which are beyond the scope of this chapter.

Note

1 The name of the island and the company implementing the restoration project are purposefully omitted in order to protect the identity of the communities involved and because the project is currently ongoing. Project statements are thus not directly referenced, but all information is taken directly from project media and publications.

References

Agrawal, Arun and Redford, Kent. 2009. Conservation and displacement: an overview. *Conservation and Society* 7 (1): 4–16.
Barbesgaard, Mads. 2018. Blue growth: savior or ocean grabbing? *The Journal of Peasant Studies* 45(1): 130–149.

Bennett, Nathan James and Dearden, Philip. 2014. Why local people do not support conservation: community perceptions of marine protected area livelihood impacts, governance and management in Thailand. *Marine Policy 44*: 107–116.

Bennett, Nathan James, Govan, Hugh and Satterfield, Terre. 2015. Ocean grabbing. *Marine Policy 57*: 61–68.

Cernea, Michael, M. 2006. Re-examining "displacement": a redefinition of concepts in development and conservation policies. *Social Change 36*(1): 8–35.

Cole, Stroma. 2017. Water worries: an intersectional feminist political ecology of tourism and water in Labuan Bajo, Indonesia. *Annals of Tourism Research 67*: 14–24.

Corson, Catherine and MacDonald, Kenneth Ian. 2012. Enclosing the global commons: the convention on biological diversity and green grabbing. *Journal of Peasant Studies 39*(2): 263–283.

Darling, Emily S. 2014. Assessing the effect of marine reserves on household food security in Kenyan coral reef fishing communities. *PLoS One 9*(11): 1–20.

Edwards, Alisdair, ed. 2010. *Reef rehabilitation manual: coral reef targeted research & capacity building for management program*. St Lucia: Coral Reef Initiatives for the Pacific.

Fabinyi, Michael and Liu, Neng. 2016. The social context of the Chinese food system: an ethnographic study of the Beijing Seafood Market. *Sustainability 8*(244): 1–17.

Fairhead, James, Leach, Melissa and Scoones, Ian. 2012. Green grabbing: a new appropriation of nature? *The Journal of Peasant Studies 39*: 237–261.

Ferse, Sebastian, Knittweis, Leyla, Krause, Gesche, et al. 2012. Livelihoods of ornamental coral fishermen in South Sulawesi/Indonesia: implications for management. *Coastal Management 40*(5): 525–555.

Ferse, Sebastian, Glaser, Marion, Neil, Muhammad and Schwerdtner Máñez, Kathleen. 2014. To cope or to sustain? Eroding long-term sustainability in an Indonesian coral reef fishery. *Regional Environmental Change 14*: 2053–2065.

Fleury, Christian. 2013. The island/sea/territory: towards a broader and three-dimensional view of the aquapelagic assemblage. *Shima 7*(1): 1–13.

Gibson, Thomas. 2005. *And the sun pursued the moon: symbolic knowledge and traditional authority among the Makassar*. Honolulu: University of Hawai'i Press.

Glaser, Marion, Baitoningsih, Wasistini, Ferse, Sebastian, et al. 2010. Whose sustainability? Top–down participation and emergent rules in marine protected area management in Indonesia. *Marine Policy 34*(6): 1215–1225.

Gorris, Philipp. 2016. Deconstructing the reality of community-based management of marine resources in a small island context in Indonesia. *Frontiers in Marine Science 3*(120): 1–15.

Hayward, Philip. 2012. Aquapelagos and aquapelagic assemblages. *Shima 6*(1): 1–11.

Hein, Margaux, Birtles, Alistair, Willis, Bette L. et al. 2019. Coral restoration: socio-ecological perspectives of benefits and limitations. *Biological Conservation 229*: 14–25.

Kenney-Lazar, Miles 2012. Plantation rubber, land grabbing and social-property transformation in southern Laos. *The Journal of Peasant Studies 39*(3–4): 1017–1037.

Kittinger, J.N., Bambico, T.M., Minton, D. et al. 2016. Restoring ecosystems, restoring community: socioeconomic and cultural dimensions of a community-based coral reef restoration project. *Reg Environ Change 16*, 301–313.

Knaap, Gerrit and Sutherland, Heather A., eds. 2004. *Monsoon traders: ships, skippers and commodities in eighteenth-century Makassar*. Leiden: KITLV Press.

Lowe, Celia. 2013. *Wild profusion: biodiversity conservation in an Indonesian archipelago*. Princeton: Princeton University Press.

Mattulada, A. 1994. The Spermonde Archipelago, its ethnicity, social, and cultural life. *Torani 5*: 104–115.

McShane, Thomas and Wells, Michael P., eds. 2004. *Getting biodiversity projects to work: towards more effective conservation and development*. New York: Columbia University Press.

Persha, Laureb, Agrawal, Arun and Chhatre, Ashwini. 2011. Social and ecological synergy. *Science 331*: 1606–1608.

Pet-Soede, C., Van Densen, W.L.T., Hiddink, J.G. et al. 2001. Can fishermen allocate their fishing effort in space and time on the basis of their catch rates? An example from Spermonde Archipelago, SW Sulawesi, Indonesia. *Fisheries Management and Ecology 8*(1): 15–36.

Radjawali, Irendra. 2012. Examining local conservation and development: live reef food fishing in Spermonde Archipelago, Indonesia. *Revista de Gestão Costeira Integrada* 12(4): 545–557.

Reid, Anthony. 1999. *Charting the shape of early modern southeast Asia*. Chiang Mai: Silkworm Books.

Sanciangco, Jonnell C., Carpenter, Kent E., Etnoyer, Peter J. and Moretzsohn, Fabio 2013. Habitat availability and heterogeneity and the Indo-Pacific Warm Pool as predictors of marine species richness in the tropical Indo-Pacific. *PLoS One 8*(2): e56245.

Tsing, Anna Lowenhaupt. 1993. *In the realm of the Diamond Queen: marginality in an out-of-the-way place*. Princeton: Princeton University Press.

Walley, Christine J. 2010. *Rough waters: nature and development in an East African Marine Park*. Princeton: Princeton University Press.

West, Paige. 2006. *Conservation is our government now*. Durham: Duke University Press.

West, Paige. 2016. *Dispossession and the environment: rhetoric and inequality in Papua New Guinea*. New York: Columbia University Press.

6

THE FLOWER GARDEN BANKS AND THE PARAMETERS OF AQUAPELAGIC SANCTUARY

Philip Hayward

Introduction

The Flower Garden Banks (FGBs), located in the Gulf of Mexico, due south of the Texas-Louisiana border (Fig. 6.1), are protrusive diapirs – geological features caused by upwellings of salt that cause dome-shaped distortions in the strata above them. When diapirs occur close to the surface, they often protrude above flat areas of land or, if located beneath the sea, appear as sea-mounts. One aspect of diapirs that has been key to their experience during the Anthropocene is the concentration of valuable mineral resources (namely salt and oil) in their upper portions, facilitating relatively easy human exploi-tation via excavation and/or drilling. The pursuit of such extractions with-out proper safeguarding and/or remediation activities has led to a variety of damage to hollowed-out diapirs, including subsidence on the crest of their domes, internal collapses, and – for terrestrial domes such as Louisiana's five (so-called) land islands (Jefferson, Avery, Weeks Cote Blanche and Belle Isle) – damage caused by ground water seepage or, on occasion, rapid inun-dation from surrounding watercourses.[1] This experience typifies what Pugh (2018, 93, 97) has termed "the shifting stakes of the Anthropocene" that have delivered "dynamically interweaving spatial relations". This chapter examines the particular assemblage that has been generated around the FGBs, first by fishing and, more recently, by the restriction of fishing and the management of the ocean floor and the waters above it as a National Marine Sanctuary (NMS). In considering the nature of the sanctuary established around the FGBs, I pay particular attention to how we can perceive, visual-ize and comprehend such assemblages within a variety of discourses. This

DOI: 10.4324/9781003569534-6

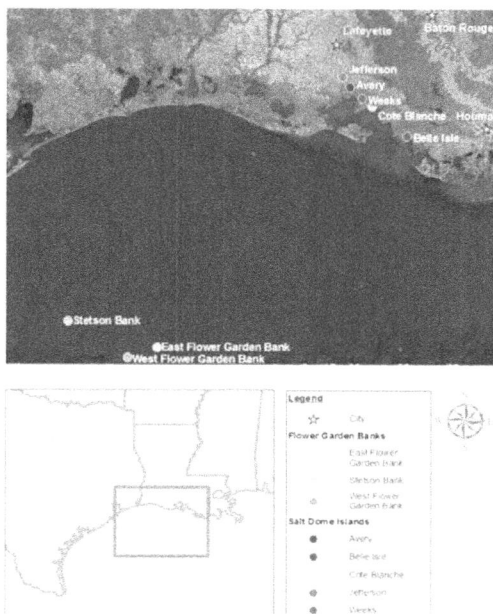

FIGURE 6.1   Location of the Flower Garden Banks and of Stetson Bank, and Louisiana's five terrestrial salt dome islands (Jefferson, Avery, Weeks, Cote Blanche and Belle) and adjacent cities. (Alahna Michele Moore 2017)

emphasizes the manner in which aquapelagos are both created by human activities of various kinds and imagined-into-being in various ways.

In what follows, I draw on frameworks developed by Hawkins (2018) with regard to geopolitical discourses and submarine and subterranean spaces. Hawkins develops her analyses with regard to Ferdinandea/Graham Island, a volcanic isolate that briefly rose above the surface of the Mediterranean, southwest of Sicily, in 1831. Noting that "subterranean and submarine spaces often attract attention in part through their resistance to or complication of regimes of visualisation", Hawkins emphasizes the need for an aesthetic approach to the "regimes of visualisation involved" that actively "enables a complication of thinking around how it is we know subterranean spaces, their dimensions, dynamics and materialities, all features important to their geopolitical significance" (2018, 1). Like her discussion, this chapter also points to the manner in which such analyses contribute to "recent geopolitical engagements with the complexities of the volumetric and the vertical, of terrain and territory, and the geopolitical and geophysical" (Hawkins 2018, 1). As she succinctly states, one of the key challenges for aesthetic and epistemological inquiry is "how exactly it is that we can come to know these spaces, and hence render them legible and controllable" (2018, 6). In addition to the visually oriented framework and discussions offered by Hawkins, this chapter considers the terminologies used to refer to subaquatic features, examines the wording of US legislation concerning them, and the extent to which concepts of "flowers" and "gardens" have been crucial to perceptions of the FGBs' space, place and significance (in contrast to those of the [oil] "fields" that surround them). This discussion, in turn, leads to a recognition of the limited security and effectivity of marine reserves with specific regard to the nature of the "sanctuary" invoked in their designation as NMSs. The chapter thereby investigates how we conceive of aquatic spaces and/or habitat islands through representation. In mounting this analysis, the chapter draws on the interrelated concepts of *shima* and the aquapelago to examine NMSs, in general, and the FGB NMS, in particular, and points to the usefulness of the two concepts for wider research on and conceptualization of oceanic environments in the Anthropocene. Similarly, the case study of the FGBs contributes to an understanding of such features as existing within highly compromised local-level contexts. Through these considerations, the chapter addresses questions of scale and of the complexity of terrestrial- oceanic relations in the Anthropocene that are significant for humans seeking to intervene in them for preventive and/or remedial purposes.

The Flower Garden Banks: Location, Structures and Human Discovery

> Salt domes and associated cap rocks and sedimentary rock hosts provide one of the most diverse economic entities of any geologic feature.
>
> *(Posey & Kyle 1988, 8)*

There are estimated to be over 500 diapirs of various sizes and elevations protruding from the seafloor on the continental shelf around the Gulf of Mexico (an arc that extends from Mexico's Yucatán Peninsula, in the south, to Florida, in the northwest), together with a number of larger, subsurface formations known as salt massifs (Posey & Kyle 1988). Salt strata formations around the coast and on its adjacent continental shelf have been widely drilled for oil over the last century and the discovery of protrusive salt dome formations on the seafloor has often been accompanied by densely clustered offshore drilling operations. Along with indicating the likely presence of oil reserves below their caps, protrusive submarine diapirs have also attracted human attention by providing raised areas that facilitate clusters of distinct aquatic vegetation, coral, fish and crustaceans. The elevated formations constitute biogeographical entities of a type referred to by Whittaker and Fernandez-Palacios (2007, 11) as "marine habitat islands" within an otherwise "hostile matrix". The "matrix" referred to by the latter is that of the surrounding terrain and environments but may, in the context of this chapter, also be extended to refer to the "hostile matrix" constituted by both intrusive fisheries and industrial oil extraction operations across the Gulf.

The East and West FGBs, together with the adjacent Stetson Bank (which has been associated with the former two through designated protection measures) occur within a dense cluster of diapirs located off the coastline between Texas and northwestern Florida. Despite the prolonged history of habitation of the Gulf Coast region (dating back to the late Pleistocene Era), and the presence of coastal peoples around the coast of the (present-day) states of Texas and Louisiana at the time of European "discovery", invasion and settlement in the 1500s, there is no evidence of any social awareness of the seafloor features discussed in this chapter prior to the mid-20th century. This, in itself, does not establish a lack of prior knowledge of the features by previous inhabitants of the region, but, similarly, there is no evidence that pre-European coastal populations fished as far as 200 kilometers offshore, let alone in the area in which the FGBs are located. In this sense, the FGBs' alleged "discovery" by US snapper and grouper fishermen in the late 1800s – albeit through the proxy of the availability of clusters of fish not present elsewhere in the deep waters off the continental shelf – may be an actual one (in human terms at least). The shallowness of the waters and the richness of the biota within them facilitated an aquapelagic assemblage that comprised the surfaces of the banks, which were dredge fished, the shallow waters above them, which were fished with nets and lines, and the surface upon which the fishing boats floated.

The mapping of seafloor features in the Gulf of Mexico is largely a 20th- and 21st-century phenomenon that has been undertaken by survey vessels of various kinds using a range of depth-measuring technologies. The FGBs are located on an area of continental shelf that gently declines from a depth of 100 meters, to the FGBs' immediate north, to 120 meters, to their immediate south. The coral-covered peaks of the FGBs rise to around 40 meters below the ocean surface, crowning masses that cover seafloor areas of c55 square kilometers (West Bank) and c75 square kilometers (East). While fishermen appear to have known that an area of sea in the vicinity hosted grouper and snapper usually found in shallower waters from the late 1800s on, the FGBs were identified and named only in the early 1900s. At this time, fishermen established their existence and named them after the vividly colored corals that lay close to the surface on their peaks, samples of which were occasionally dragged up in fishing nets. The terminology they used is significant, emphasizing terrestrial perceptions of an aquatic space. The word "flowers" – rarely used by marine botanists (at least, with regard to subaquatic plants) – can be seen as a favorably analogous characterization of coral to terrestrial plants. Similarly, the term "garden" is analogous – but in a more complex manner. Its everyday usage refers to an area in which decorative or food plants are cultivated, but the term also has a prominent biblical association in the form of the Old Testament's Garden of Eden. The latter has two primary associations, first as a bucolic locale amply blessed with fruits of all kinds and, second, as an ideal space irretrievably violated by human folly in breaching prohibitions on their use of the space. Adding to these associations, the term "garden" also has a distinct meaning in American English, referring to a large bounded recreational space (such as New York's Madison Square Garden arena) (Pollack 2010). As subsequent discussions will elaborate, the aggregation of these conceptual aspects in the FGBs' name is far from incidental to their subsequent imagination, representation and elaboration. Note, for instance, the National Oceanic and Atmospheric Administration and Sea Turtle Conservancy's 2001 project webpage (now defunct[2]). After comparing diapirs to "underwater gardens", the page described FGBs as "like oases in the desert", a characterization that was simultaneously vivid (emphasizing their biological richness), biogeographically apposite and is also evocative of various biblical oases (such as Elim, featured in Exodus 15:27). The banks' name and related characterizations provide them with an enduring associative "charge" which informs perception and discussion of them in a manner that, for instance, the names of other Gulf banks – such as Geyer, Rankin or Sonnier – do not.

Knowledge of the FGBs appears to have remained limited for much of the early to mid-20th century. The US Navy and American Society for Oceanography conducted surveys around the FGBs in 1968, mapping their shape and the topography of the seafloor around them through depth-sensing technologies. As Hawkins (2018, 22) has identified, our current perception and mapping of such features differs from traditional cartography in that the former "has been premised on seeing" whereas subsurface oceanic features are comprehended and represented through "multisensory" media. Noting that "many visualisations of the sea bed are actually formed through data produced by practices of sounding", she reflects that "what emerges". within a geo-strategic framework, "is a need for more multisensory ways of knowing as part of how we both understand and mobilise volumetric practices of power". With regard to Hawkins's discussions, it is worth reiterating that the first perception of the existence of the FGBs was produced by proxy, in the form of fish clustered around particular points in the sea that *suggested* the existence of subsurface peaks. This is a significant point. The fish inhabited a space *around* the banks that was dependent/contingent on them but was different in substance, mass and character from them. The colorful coral fragments that were dredged up were more closely aligned to the materiality of the banks – in that they grew *upon* them – but were, nevertheless, different in form from the banks themselves and, therefore, can be considered just as much a proxy of the FGBs' existence as the fish were. So, from their earliest recognition by humans, the FGBs were cognitively brought into being through a convergence of indicators, and they were geomorphic formations substantially defined by the plant and coral types that adhered to their upper surfaces and of the fish that swam around them. In these ways – and particularly through their linguistic designation as gardens – the FGBs are as much rhetorical constructs as they are material entities.

Discussing the short-lived manifestation of Ferdinandea/Graham Island, Hawkins invokes the aquapelago as "a land-oceanic continuum that when mobilized requires that we resist any easy distinction between the subterranean and submarine" and contends that the concept problematizes "politico-legal concepts of territory" premised on notions of fixed terrestrial spaces (Hawkins 2018, 3, 10). This characterization is equally apposite for the FGBs' comprehension and representation by human agencies – and to subsequent eco- and geo-political discourses, strategies and legislative measures passed to restrict human use of its subsurface spaces – since it emphasizes the FGBs and NMSs as entities *performed by* an intersection of animate and inanimate objects and the abstract discourses applied to them. In the case of the FGBs NMS, language has been a key element of that performance.

Prior to European settlement and 20th-century development of the shore and hinterlands of the Gulf of Mexico, the waters off its immediate coast supported a rich biomass that was a significant resource for the coastal populations that harvested various organisms in its coastal shallows. Various types of marine vessels, fishing gear and harvesting techniques introduced by European settlers extended the scale and extent of the fisheries and created a close involvement of coastal populations with in-shore waters. From the late 1940s on, this offshore activity was accompanied by another – the extraction of oil from offshore deposits located in salt strata around the continental shelf. It is notable, with regard to the discussions advanced in this chapter, that the term used to refer to both terrestrial and seafloor subsurface oil deposits is "field". This term derives from the Old English term *feld*, which originally referred to a tract of open land and was later used to refer to "a parcel of land marked off and used for pasture or tillage" (Etymology Online n.d.). The term subsequently became extended to more specific areas, such as "sports fields" in the late 1800s (Etymology Online n.d.). The first use of the term "oil field" appears to have been around the 1860s (Merriam Webster n.d.) following the development of commercial extraction processes in Pennsylvania. The date of the first use of the term to describe offshore oil deposits is unclear but appears to have followed the development of sites in the Gulf of Mexico in the late 1940s. But whereas the harvesting of planted fields involves the cyclic gathering of a renewable bio-resource that is maintained by humans for that purpose, the extraction of oil is more accurately a (one-off) draining and transfer of a preexistent geomorphologic resource without remediation of the local environment. In this sense, the use of the term "field" by the oil industry can be seen as a cosmetic one that masks the face of one activity with a term appropriated from another context. For much of human history, agriculture has been (and has been regarded as) a livelihood activity integrated with various aspects of the natural world – with humans being one aspect of the multi-species terrestrial assemblages involved in the activity in a relatively sustainable manner. Oil extraction, by contrast, has been conducted in subsurface strata in manners that have exemplified Anthropocene degradation of the seafloor, oceanic chemistry and related biomasses.

The Gulf's offshore oil industry is premised on maritime rig technologies: metal structures connected to the seafloor which support the insertion of drill heads into the salt strata and the subsequent extraction of oil to the surface via pipes. This activity, and the related transfer of oil to ships and/or piping to hub facilities for transit to onshore processing facilities, has involved multiple disruptions of the seafloor and incidental operational pollution of seawater. This emphasizes that livelihood activities that create aquapelagic assemblages in particular coastal and aquatic locales are frequently disruptive to the broader ecological health and stability of those locales. But while the ongoing leakage of pollutants is a continuing problem, there has been equal concern over the increasing pollution of Gulf waters by fertilizer runoff from onshore agribusiness operations that have seeped into the Mississippi and its tributaries. This runoff is deposited into the ocean via the Mississippi Delta, creating a hypoxic (i.e., oxygen-depleted) zone just above the seafloor which extends for around 15,500 square kilometers around the shore of the Gulf. While the pollution spikes and subsides in an annual cycle, dependent on patterns of application of fertilizers, its peak periods massively diminish demersal (i.e., bottom- dwelling) fish and shellfish stocks in a continuing cycle that impedes breeding and causes "dead zones" (Diaz & Rosenberg 2008).

In 2010, the catastrophic condition discussed above was compounded by the most sudden, intensive and visibly perceptible pollution event caused by the offshore oil industry to date: the massive spillage of oil from the offshore Deepwater Horizon (DH) rig. DH was conducting exploratory drilling in 1,600 meters of water in the Macondo prospect zone, some 66 kilometers off the southeastern Louisiana coast (within the offshore oil drilling areas identified in Fig. 2). The DH spillage was triggered by a rapid upflow of methane gas from the subsurface salt strata to the drillhead platform, where it ignited in an explosion that severely damaged its wellhead in a manner that allowed the oil deposit to gush unimpeded into the ocean from the drill bore. This forceful leakage injected around 780,000 cubic meters of oil into the Gulf (Flow Rate Technical Group 2011) in the form of both surface slicks and submarine plumes, causing both immediate and short- to medium-term fatalities in bird, marine mammal and demersal fish and shellfish species in the Gulf and along adjacent shorelines (Wallace et al. 2017). As should be evident, the inadvertently engineered injection of a large volume of viscous fluid into the saline hydrology of the Gulf during the DH incident proceeded without reference to any legislated (or otherwise assumed) spatial delineations of "protected" areas of the Gulf and/or its ocean floor. This aspect is particularly pertinent with regard to the following aspect of this chapter's discussion of the FGBs, their designation as an NMS.

National Marine Sanctuaries

NMSs are distinct from terrestrial environmental reserves in that the latter are almost exclusively defined by borderlines that demark particular land areas from their surrounds (and do not attempt to delineate areas of airspace above them or, with few exceptions, subterranean spaces beneath them). NMSs, by contrast, operate as demarcated areas that comprise the volumes of water that occupy the spaces between cartographic reference points and include the seafloors of these areas. Such "sanctuaries" are usually associated with waters around islands or coastal locations or else located above shallow features (such as sub-surface plateaux, seamounts, reefs, etc.) that support marine species that are effectively tethered to the submarine features and their shallows.

The US National Marine Sanctuary System was created under Title III of the US Marine Protection, Research, and Sanctuaries Act (1972) (henceforth referred to as "the Act") (NMS 2015). The use of the term "sanctuaries" in this context merits attention. The first significant aspect is that the latter is not one of the 13 terms defined in the preamble to the Act (despite its obvious significance as one of the three key terms in its title). In everyday usage the term "sanctuary" has three principal meanings (the latter of which derive from the former). Etymologically, the term derives from the Latin word *sanctus*, meaning holy, with sanctuary referring to a holy site, such as a consecrated Christian place of worship of the type that has allowed fugitives of various types to "seek sanctuary" – i.e., safe (if largely temporary) domicile – within them. This usage, in turn, resulted in the term being used more generally to refer to spaces created under various ordinances that offered safe domicile to other species threatened by humans (and/or other species), giving us the concept of the "wildlife sanctuary". While it would seem apposite to understand the use of the term "sanctuary" in the US Act primarily in the latter sense, it is not inappropriate to also perceive it to be imbued by a spiritual sense of holiness transposed to a secular context (in terms of environmental values). But there are also more complex levels at work in the ascription of sanctuary status to assemblages of animate and inanimate elements. Humans and human perceptions and values play a key role here, both as co-creators of such assemblages and as objectifiers of them. While his development of these concepts has been hitherto little remarked upon by scholars, a number of publications by Japanese anthropologist and Island Studies theorist Jun'ichiro Suwa are particularly pertinent for the above considerations.

Suwa's initial work on the concept of *shima*, particularly as it is understood in Japan's Amami Islands, identified the manner in which a term that is usually translated into English as "island" is far more nuanced, embodying a dual meaning: "islands as geographical features and islands as small-scale social groups where cultural interactions are densely intermeshed" (Suwa 2007, 6). As Suwa elaborates, in traditional Amami culture, each *shima* is "enacted through various practices and performances of demarcation" and is "a work of territorial imagination, an extension of personhood and a 'cultural landscape'" (Suwa 2007, 6). In this sense, he contends:

> a shima is a sanctuary, in that the natural environment and social space are articulated in such a way that one imagines them as a totality. ... Territory is thereby strongly associated with the livelihood of communal social spaces and a particular type of territorial imagination, for its boundary consists of landmarks. Waters, coastlines and hills mark shima; the territories are imagined and created as a place ... produced through boundary marking ravines, reefs, or hills, providing resources for imagination and the drawing of lines and shapes on landscapes.
>
> *(Suwa 2007, 6)*

In a subsequent article that discusses the compatibility of concepts of *shima* with the concept of the aquapelago, he contends that:

> One of the possible ways of grasping the nature of aquapelagic assemblages might be to focus on locations where fractal concentric circles initiate activities. This point might be regarded as a "sanctuary", using an expanded definition of that term that includes its complementary characterisations of a holy/spiritual centre, a place of refuge and safety and a reserve where flora and fauna are protected.
>
> *(Suwa 2012, 14)*

Noting his indebtedness to Akimichi's (2004) concept of *kami no commons* (which derives from the type of sanctuary offered by Shinto shrines), Suwa contends that:

> Aquapelagic assemblages can be conceived as sanctuaries with regard to either real or imagined space. ... Key to the imagination of sanctuary is a shared idea that constructs cultural reality and thereby becomes crucial to any such project. Sanctuary therefore is a type of spatio-temporal space where things take place, operate and interact within a particular framework. ... Aquapelagic assemblages—and/or the sanctuaries they comprise—are not necessarily self-contained, self-sufficient and/or self-sustainable. They are not bounded, ahistorical "utopias" [of] the stereotypical "idyllic" or "pastoral" landscape. ... Aquapelagic assemblages are, rather, specific products of ongoing processes in actual locations. ... [As] sanctuaries manifested in land/seascapes, they are occupied

with sacred, untouchable, memorised and/or identified elements that are shared through and as common knowledge, skill, consciousness, desire and ideology.

(Suwa 2012, 14-15)

This sophisticated characterization is entirely apposite to the discussion of NMSs in general and the FGBs NMS in particular. Indeed, Suwa's contention summarizes a set of perceptions that illuminate the concept of sanctuary embodied with the Act (and its subsequent modifications), various assertions of it, and the rationales for attempts to gain NMS status for particular locations. My assertion of this in no way suggests that the authors of the Act consciously conceptualized NMSs in such manners but, rather, that the framework of the Act and the various clauses that create NMSs as legislated entities involve a similar process of recognition-and-becoming that Suwa ascribes as an organic aspect of human imagination and management of space.

While the original US Marine Protection, Research, and Sanctuaries Act, passed in 1972, addressed the establishment of NMSs as something of an afterthought to its primary focus on regulating the dumping of harmful materials in coastal waters (in its Clauses I–II); Title IIIa Finding 2 of the modified version of the Act passed by Congress in 1984 contained a significant expansion of the factors that might lead to designation of an area as an NMS, identifying that:

certain areas of the marine environment possess conservation, recreational, ecological, historical, scientific, educational, cultural, archeological, or esthetic qualities which give them special national, and in some cases international, significance.

(US Government 1972)

Finding 4 identified that the Federal program established by the Act aimed to:

a. improve the conservation, understanding, management, and wise and sustainable use of marine resources;
b. enhance public awareness, understanding, and appreciation of the marine environment; and
c. maintain for future generations the habitat, and ecological services, of the natural assemblage of living resources that inhabit these areas.

(US Government 1972)

The identification of a number of human activities and perceptions as providing national and international "significance" to specific marine environment sites and the reference to "natural assemblages" appears to imply an inter-face between the two that creates an implicit recognition of the aquapelagic nature of the sites, i.e., their establishment as nationally significant as a result of aquapelagic interactions around them and perceptions of them. Rather than simply authorizing the establishment of aquatic wilderness zones in which non-human biota can be preserved with minimal human intervention, the Act explicitly emphasized both the nature of particular marine areas as *shima* (in the sense elaborated by Suwa above) and their aquapelagic aspects (and the ability for these to be maintained) as key aspects of the designation of NMSs. While such designation (to date) precludes commercial fishing and blocks oil prospecting and extraction within sanctuary boundaries, socio-economic concerns such as recreation (including recreational fishing) and tourism have always been pivotal to the designation.

But before moving to analyze particular sites, it is worth returning to the principal purpose of the first iteration of the Act: i.e., to minimize dumping in and related pollution of US marine areas. While the provisions of the Act had some impact on industrial practices, the extent and impact of the DH oil spill in the Gulf of Mexico in 2010 demonstrated that existing protec-tions and penalties were woefully inadequate. Responding to research into industry practices, President Barack Obama established the Bureau of Safety and Environmental Enforcement in 2016 to monitor and enforce new guide-lines on oil industry operational safety at a level never previously seen in the United States. This enterprise was, however, short-lived. Shortly after gain-ing office in 2017, President Donald Trump signed an executive order enti-tled "The America First Energy Strategy", began winding back the Bureau's powers and handed much of the control back to (predominantly Republican controlled) states in which the oil lobby exercises significant influence, and President Biden's presidency has yet to regain the lost ground on this.

Designated Sites

The first NMS was established in 1976, four years after the Act was passed (with the responsibility for approval having shifted from the Secretary of Commerce to the President in the interim). The first sanctuary was a maritime heritage site designated by President Gerald Ford with somewhat loose adherence to the provisions of the first iteration of the Act (being an area off the North Carolina coast where the sunken Civil War–era ship the USS *Monitor* had been discovered in 1972). A second site approved by Ford later that year was of a significantly different type. The Key Largo NMS was established to protect the only living coral barrier reef on the US East Coast, which was – and remains – a major dive attraction for tour-ists. Florida's nomination of the site was an early indication of the state's intent to maintain and develop its offshore waters for tourism purposes (in marked contrast to its systematic devastation of its Everglades swamp system [see Grunwald 2007]). This intent also led the state to prohibit offshore oil drilling through various measures in the following decades, including a statewide ban approved in a state referendum held in association with the US mid-term primaries in November 2018.

The replacement of Republican president Ford by Democrat Jimmy Carter in 1977 led to a new impetus for establishing NMSs with the Department of Commerce's National Oceanic and Atmospheric Administration (NOAA) drawing up a list of 67 potential candidate areas. Carter went on to authorize the establishment of four NMSs, around California's Channel Islands (1980) and California's Point-Reyes Farallon Islands, Florida's Looe Key and Georgia's Gray's Reef (all in 1981). Following a period of inactivity during the first four years of Ronald Reagan's presidency, NOAA was successful in gaining approval for the establishment of an NMS in Fagatale Bay in American Samoa in 1985. Further NMSs were approved off the Florida and California coasts in 1989–1990, following President George Bush's inauguration, before a 13-year-long campaign to have the FGBs designated succeeded in 1992. In 1996 the nearby Stetson Bank, a "sandstone bank with fire coral and sponge-covered pinnacles and flats" (NOAA n.d.), was added to the FGB NMS, with the three being jointly administered and represented by NOAA's National Ocean Service (Fig. 6.2). Further expansion

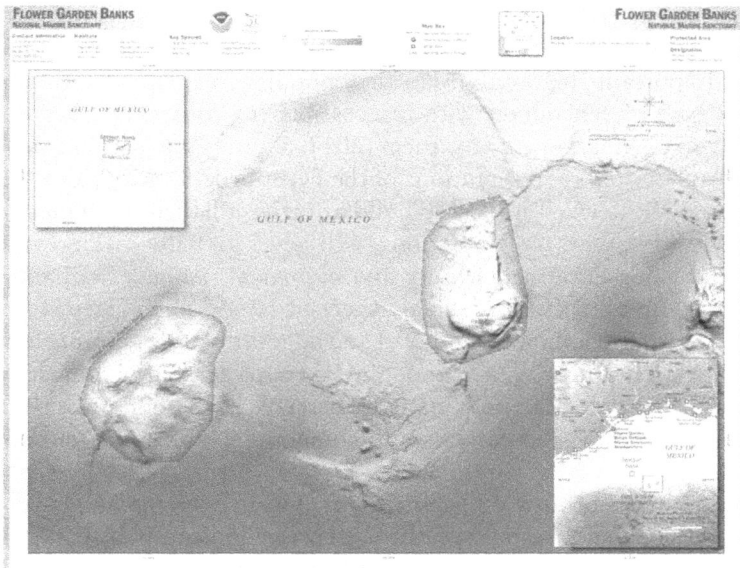

FIGURE 6.2   Contour map of the original Flower Garden Banks National Maritime Sanctuary (delineated by red perimeter lines), with inset showing position of Stetson Bank. *Source:* NOAA/public domain (n.d)

followed in 2021 when 14 further reefs and banks were added, expanding the designated area to 414.5 square kilometers (NOAA 2021).[3] NOAA has identified the FGB NMS's "key habitats" as "sand flats, soft sediments, bank reefs, drowned reefs, pinnacles, hard substrate … algal sponge communities, brine seeps/flows, fault scarps, and artificial reefs" and has identified significant species as including "star coral, brain coral, manta rays, hammerhead sharks, and endangered loggerhead sea turtles" (NOAA n.d.). These, together with the various humans who visit and/or otherwise engage with the area through its representations, can be considered to constitute the FGBs NMS's general aquapelagic assemblage.

As entities located a considerable distance from the coast and at depths not visible to surface explorers using glass-bottomed boats, snorkels, etc., the raised seafloor features and the flora and fauna congregating around them are currently directly accessible only to those divers who can arrange their transport offshore and dive down to them wearing breathing apparatus. More generally, the FGBs are principally manifest to and apprehensible by the public through their representation via media products on NOAA's website pages devoted to the FGBs NMS. These include a series of maps, including the most commonly used image (Fig. 3), and a webpage devoted to the FGBs. The (five-episode) *Sanctuary Video Series* (NOAA 2001) also gives viewers lovingly assembled "diver's-eye" perspectives on the colorful marine life on and around the Banks in a manner that derives from Jacques Cousteau's pioneering documentaries in the 1950s and persists in contemporary dive documentaries. As Mikkola (2018) emphasizes, such productions have an important function within the "popular imaginary", providing us with the type of "encounter with inaccessible environments" that we need in order *"to be able to care"* for them (2018, 5 – my emphasis). In this manner, the audiovisual representations of the FGBs produced by NOAA can be perceived as PR devices utilized to build affective alliances with members of the public, who, with few exceptions, will never actually dive down to the Banks and experience their beauty and biological richness firsthand. Such public relations activity is a crucial aspect of NOAA's management of and advocacy for NMSs. Public interest in and support for the continued protection and designation of NMSs cannot be taken for granted, but must, rather, be nurtured in order for politicians to perceive such support and to take that into account in their responses to competing commercial and political agendas.

In the manner outlined above, the FGBs exist in overlapping planes. They exist as entities that are apparent through proxy indicators (fish, coral, etc.) that are spatially appraised and visually rendered through depth-sensing technologies and are represented in audiovisual media deployed by divers. They also exist as the focal components of a legislated entity that delineates and articulates the space around them as an NMS. The FGB NMS is, thereby, a representative spatial instantiation of the legislative-rhetorical discourse and prohibition mechanisms of the NMS program (embodying all the complexities of the latter) as much as it is a place with an objective existence.

Conclusion

Commencing with discussions about the conception and nomenclature of seafloor spaces, this chapter has sought to engage with Pugh's (2018, 99) apposite and timely observation that "thinking about islands in terms of multiple, unfolding temporalities and the richness of relationality in the Anthropocene raises some new questions for island ontology" and, in particular, his attempts to push the bounds of "relationality" of islands into consideration of "TARDIS-like" spaces (Morton 2016) where totalities are "greater than anything which can be experienced phenomenologically" (Pugh 2018, 100). The maps, website and video representations of the FGBs' subaquatic flora and fauna, and ongoing data gathering on biota, water conditions, etc., by NOAA, are assembled and displayed to assert, manifest and inform the maintenance of the FGBs NMS as a *defended* entity. In this regard, they typify – in *micro* – what Springer and Turpin (2017, 18) refer to as the "morphology of sovereignty", which increasingly "exhibits" itself through "dematerializations, virtualizations, physical reassertions, and material instantiations". The FGB NMS might thereby be seen to project and aspire to a relational "micro-sovereignty" within the "hostile matrix" of the Gulf of Mexico that is as rhetorical as the micronational statuses that have been asserted for various islands and offshore platforms in various locations over the last century (*Shima* 2023). Located within the globally integrated space and immediate temporality of the Anthropocene, and immersed in a liquid environment that cannot support any protective membrane other than the highly permeable and fragile provisions of national legislation, the FGBs' sanctuaries resemble Ferdinandea/Graham Island, as discussed by Hawkins (2018). They have risen and been subject to claims and contestation at a particular historical juncture and are as fragile in temporal terms as they are in environmental ones. In particular, the fragility of environmental regulations in the current polarized era of US politics gives little cause for optimism. Just as aquapelagos, in general, wax and wane as climate and environmental conditions and related human uses of space alter, so, too, does the effectiveness of the protection of marine areas prescribed by sovereign governments at particular historical moments. Sanctuary, whilst precious, is usually fleeting.

Acknowledgements

Preliminary research for this chapter was conducted as part of a visiting research fellowship at the Midlo Center for New Orleans Studies at the University of New Orleans in 2016. Thanks to Connie Zeanah Atkinson from the Midlo Center for facilitating my visit; to Kelly Drinnen from NOAA for assisting with my research inquiries in 2017–2018; to Alahna Michele Moore for her research assistance and mapping skills; to Jun'ichiro Suwa for both inspiring many lines of thought and for providing feedback on an earlier draft. The first version of this chapter was written in the convivial environment of the University of Hong Kong in 2019; many thanks to Otto Heim for facilitating this, and to Amelia Coyle-Hayward for giving me her perspective on my early sketches.

Notes

1 See Hayward and Moore (2018) for discussion of the differing fates of the land islands over the last fifty years.
2 The site was formerly available via htts://doi.org./10.5962/bhl.title.11363.
3 The only NMS added during the Trump presidency was the Mallows Bay-Potomac River area in Maryland, which is the site of over 100 sunken ships dating from the Civil War to World War I.

References

Akimichi, Tomoya. 2004. *Commons no jinruigaku*. Tokyo: Jinbun Shoin.
Diaz, Robert J. and Rosenberg, Rutger. 2008. Spreading dead zones and consequences for marine ecosystems. *Science, 321*(5891), 926–929.
Etymology Online. n.d. Field. https://www.etymonline.com/word/field
Flow Rate Technical Group. 2011. *Assessment of flow rates estimates for the Deepwater Horizon/Macondo Well oil spill*. National Incident Command, Interagency Solutions Group.
Grunwald, Michael. 2007. *The Swamp: The Everglades, Florida and the politics of paradise*. New York: Simon and Schuster.
Hawkins, Harriet. 2018. A volcanic incident': Towards a geopolitical aesthetics of the subterranean. *Geopolitics, 24*(3), 1–26.
Hayward, Philip and Moore, Alahna Michele. 2018. Vertical features in flux: Elevation, interiority and the Anthropocene disruption of South West Louisiana's five salt dome islands. *Journal of Marine and Island Cultures, 7*(2), 37–45.
MerriamWebster. n.d. Oilfield. https://www.merriam-webster.com/dictionary/oil%20field
Mikkola, Hedi. 2018. Movements beyond human: Ecological aesthetics and knowledges in underwater wildlife documentaries. *Finnish Journal for Human-Animal Studies, 4*, 4–26.
Morton, Timothy. 2016. Molten Entities. In *New Geographies 8: Island* edited by Daniel Daou and Pablo Pérez-Ramos. Cambridge: Universal Wilde, 72–76.
National Marine Sanctuaries. 2015. National Marine Sanctuary 50th anniversary timeline. https://sanctuaries.noaa.gov/about/history/
National Oceanic and Atmospheric Administration. 2021. Sanctuary expansion. https://flowergarden.noaa.gov/management/sanctuaryexpansion.html

National Oceanic and Atmospheric Administration. n.d. Sanctuary Video series. https://flowergarden.noaa.gov/image_library/videoseries.html

Pollack, Michael. 2010, July 9. Why is Madison Square Garden called a garden? Was it ever a garden? *New York Times.* https://www.nytimes.com/2010/07/11/nyregion/11fyi.html

Posey, Harry H. and Kyle, J. Richard 1988. Fluid-Rock interactions in the salt dome environment: An introduction and review. *Chemical Geology,* 74, 1–24.

Pugh, Jonathan. 2018. Relationality and island studies in the Anthropocene. *Island Studies Journal, 13*(2), 93–110.

Shima. 2023. Micronationality anthology. https://shimajournal.org/anthologies/micronationality.php#gsc.tab=0

Springer, Anne Sophie and Turpin, Etienne. 2017. The Science of letters in *Reverse hallucinations in the archipelago (Intercalations #3)*, edited by Springer, Anne Sophie and Etienne Turpin. Berlin: SYNAPSE—The International Curators' Network, K. Verlag & Haus der Kulturen der Welt, 1–52.

Suwa, Jun'ichiro. 2007. The space of shima. *Shima, 1*(1), 1–14.

Suwa, Jun'ichiro. 2012. Shima and aquapelagic assemblages: A commentary from Japan. *Shima, 6*(1), 12–16.

United States Government. 1972. US Marine Protection, Research, and Sanctuaries Act. https://web.archive.org/web/20120508220037; http://www.epa.gov/about epa/histo ry/topics/mprsa/

Wallace, Bryan P., Brosnan, Tom, McLamb, Danya, et al. 2017. Effects of the Deepwater Horizon oil spill on protected marine species. *Endangered Species, 33,* 1–7.

Whittaker, Robert J. and Fernandez-Palacios, José María, eds. 2007. *Island Biogeography: Ecology, evolution and conservation.* Oxford: Oxford University Press.

7

THE JUAN FERNÁNDEZ ISLANDS IN TRANSITION

Cruise Tourism, the Commodification of Nature and the Establishment of a National Park

Elizabeth Chant and Natalia Gándara Chacana

Introduction

Scholars have previously identified the role of islands as sites where fantasies are projected of idealized, idiosyncratic and untouched forms of life and living (Baldacchino 2012, 55). Located 670 km from the South American Pacific rim, the Juan Fernández Islands (Fig. 7.1), nowadays part of Chile's V Region of Valparaíso, have stimulated the imaginations of generations of writers on account of their being a hideout for European pirates and privateers during the early modern period, recounted most famously in Daniel Defoe's classic novel *Robinson Crusoe* (1719). The archipelago consists of three main islands, Robinson Crusoe Island (formerly known as Masatierra), Alejandro Selkirk Island (formerly known as Masafuera) and Santa Clara Island. Of these, only Robinson Crusoe and Alejandro Selkirk are currently inhabited, with the primary settlement being the port town of San Juan Bautista on Robinson Crusoe. The projected population of the islands in 2021 was 1,053 people (Biblioteca del Congreso Nacional de Chile 2020).

The islands' high levels of endemism, combined with their distance from the South American mainland, and their relatively late occupation by humans, have made Juan Fernández a receptacle for idealized visions of nature untainted by human presence. With the advent of steamships and increasing demand for leisure tourism into the 20th century, from 1922 Juan Fernández became a sporadic port of call on cruises around South America. The archipelago's native flora and fauna provided unique souvenirs, and its connection to an English literary classic made it popular with Chileans, US tourists and Europeans alike. In the materials promoting these trips, and published accounts of tourist visits, we encounter an active romanticization

DOI: 10.4324/9781003569534-7

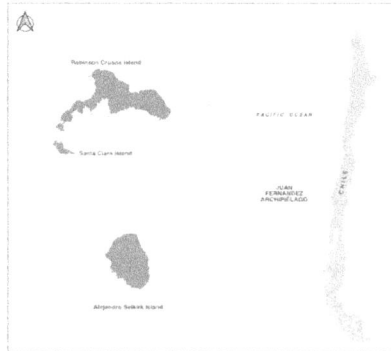

FIGURE 7.1   Map of the Juan Fernández Archipelago and its position with regard to the Chilean mainland.

of the islands' past as a buccaneers' hideaway. In this chapter, we examine how the arrival of leisure tourism in Juan Fernández has introduced the islands into new commercial networks and obscured the inhabitants' relational practices in the context of an aquapelagic society (Hayward 2012a). We further underline how the connections brought about by cruise ships have inserted the Juan Fernández archipelago into new and reconfigured transnational and transoceanic relationships. In this panorama, we therefore trace the role of the archipelago as a "malleable platform for the practice of some form of exclusivity that needs to be protected from mainland interaction or contamination" (Baldacchino 2012, 57); in this case, the islands' "unspoilt" nature.

Prior to tourism, from the late 19th century, fishing was the most important economic activity conducted by the islanders. In particular, fishing of the Juan Fernández rock lobster (*Jasus frontalis*) was the backbone of the islands' economy. By 1949, around 400 people lived on Robinson Crusoe Island, and around 20–30 people lived on Alejandro Selkirk Island (Castedo 1949a, 54). The local population today is a mixture of Chilean and European settlers, as well as descendants of penal colony families who were sent to the island both during the late 18th century and by the Chilean government after independence in the 19th century. As we highlight, their relationship with the surrounding sea and their telluric connection with the islands' landscape converged in a unique aquapelagic assemblage, one that has altered in response to changing environmental pressures.

In line with John Gillis's concept of "ecotones", coastal environments "where two ecosystems overlap" (2012, 9), Juan Fernández islanders live at the "porous and connective" interface between land and sea (Gillis 2014, 156). Having transitioned from settlers to an "endemic" population (Brinck 2007), their way of life fosters a close relationship with the ocean, understanding the sea as a projection of their lived space. Moreover, Juan Fernández islanders engage with the ocean in both horizontal and vertical terms, using it as a connecting surface but also harvesting their main source of income and sustenance from its depths. Cruise ship tourism, however, transformed the island's connection with the South American continent, and it reconfigured the economic and cultural image of the islands. The perceived "remoteness" of the islands became their key selling point in Chilean and international tourist discourses as a vestige of nature uncorrupted by mankind that featured species such as the Chonta palm (*Juania australis*) that could not be found elsewhere in South America. As such, the vision of Juan Fernández constructed "on the continent" came to predominate in literary and visual depictions of the region from the first half of the 20th century (Carmona Jiménez 2021, 59).

By focusing on the arrival of cruise tourism in Juan Fernández, this chapter considers how this particular form of tourism connects the islands to continental spaces and inserts them into national and international tourist imaginaries. We argue that cruise tourism facilitates a form of engagement with the island environment distinct from the islanders' aquapelagic relationality and creates new ways of exploiting and commodifying nature, developing a cultural identity for Juan Fernández that also saw the renegotiation of their place in Chilean national identity, and evidencing the "commodification of the natural world for capitalist gains" (Scribner 2021, 80). We highlight the distinct relationship the cruise ship creates with the aquatic and aquapelagic environments encountered and reaffirm the role of the sea in the development of leisure travel imaginaries that persist today. We begin by discussing the nature of Juan Fernández as an aquapelagic space, and how this understanding frames tourist interactions, while also introducing the context of early-20th-century cruise tourism in Chile. We then examine portrayals of Juan Fernández's main island primarily using travel accounts, newspaper articles and advertising, exploring how cruise ship tourism created new ways of profiting from the island's nature. This chapter concludes by addressing the Government of Chile's conservation policies in the archipelago, examining how they account for the aquapelagic relations of this environment and seek to counteract the damage done by commercial overfishing.

Tourism and Juan Fernández's Aquapelagic Environment

The concept of the aquapelago has been influential for understanding island communities' relationships with the "transitional zone between land and sea" that they inhabit (Hayward 2012a, 1). Drawing on the work of Pacific Ocean scholars who understand this body of water as a "sea of islands" (Hau'ofa 1994), the concept of the aquapelago is premised on the dynamic connections between both human and non-human entities in their island homes that continue to evolve in response to pressures imposed by forces such as globalization, imperialism and, increasingly, global heating. It is a space "generated by livelihood", as Suwa asserts, which is particularly important for understanding the role of tourism in aquapelagic space (2012, 13). Lindsay Bremner has previously used this concept to approximate the changes wrought by the introduction of tourism in the Maldives, noting how this caused the Indian Ocean to become

> not less, but more enmeshed in human affairs by becoming more economically, technologically and digitally mediated. New aquapelagic performances combining infrastructures, technologies and global imaginaries produced new species of islands, new kinds of oceans, new typologies of architecture and new kinds of humans, constantly reassembling the unstable continuum between geological, hydrological, human, animal and technological life according to the laws of value.
>
> *(Bremner 2017, 26)*

Drawing on Hayward's earlier description of the aquapelago as a "performed entity" (2012a, 6), here Bremner highlights the potential of tourism to facilitate new modes of interaction, as spaces are understood through an economic lens that privileges the experience of visitors to the space. The arrival of tourism also brings distinct imaginaries into conflict: the existing aquapelagic relationships of permanent dwellers are often overwritten in the hegemonic narratives of the visiting populations to present destinations that align with exoticized ideals, and, in the specific case of Juan Fernández, tap into the cultural imaginary of pirates and castaways. Donald Macleod has termed the process by which an island has both its national and its local identities manipulated to attract visitors "cultural realignment", noting the particularly problematic treatment of islands as vestiges of cultural "uniqueness" that are bounded and isolated by bodies of water (MacLeod 2013, 76). Such an understanding cultivates a vision of islands as "locales of desire, as platforms of paradise, as habitual sites of fascination, emotional offloading or religious pilgrimage" (Baldacchino 2012, 55), a kind of *tabula rasa* onto which the fantasies of the tourist can be projected. It is such an idealization that prevails in touristic depictions of Juan Fernández.

The history of Juan Fernández has long been defined by processes of commodification and extraction, which have impacted the islanders' aquapelagic relationships. Like the Galápagos Islands, the Juan Fernández Islands have only been colonized by humans "within the last five centuries" (Burke 2020, 60). Thought to have first been sighted by its namesake Spanish navigator Juan Fernández (c. 1539–c. 1604) in 1574, the archipelago was sporadically inhabited until the late 19th century. Fernando Venegas Espinoza and Sergio Elórtegui Francioli have previously identified two main phases of early Spanish occupation: from approx. 1574–1590, and from 1591 to 1616, when resource exploitation began in earnest (2022, 391). Logging, fishing and sealing were undertaken by Indigenous people who had been forcibly relocated to the islands from the continent (Venegas & Elórtegui 2022, 408). This early settlement was temporary, though, and the islands remained largely unoccupied following the second phase, becoming a hide-out for European pirates in the 17th and early 18th centuries. Permanent occupation did not occur until the 1750s when the Spanish Crown formally established a military colony to prevent rival European nations from col-onizing the islands and gaining access to the Pacific. The main island of Robinson Crusoe was occupied by military men and their families, as well as a small penal population.

When Chile became independent in the early 19th century, the republican government cemented the islands' place as part of Chile's territorial sover-eignty (Woodward 1969, 127). The islands were initially used as a penal colony, but in the late 19th century, the Chilean government also began to lease the islands to private enterprises, accelerating the exploitation of their marine and terrestrial environments and altering the islanders' aqua-pelagic relationship as their resources came under increased pressure. This establishment of commercial fishing was crucial in the construction of the islanders' identity, being inextricably linked with the sea and its non-human inhabitants. The islands were leased to Swiss-born entrepreneur Alfredo de Rodt (1843–1905) by the Government of Chile in 1877. Rodt established commercial lobster fishing, which was subsequently taken over by Carlos Fonck and Company of Valparaíso, which set up canning facilities and regu-lar shipping of products to the port city. Writing in 1904, US travel writer Marie Robinson Wright hinted at the promising future of industry in Juan Fernández: "it may some day hold as high a place in the practical world of business as it now holds in the imaginary realm of fiction" (Robinson Wright 1904, 429), although canned fish was not the only significant export prod-uct for the early Fernándezian economy.

Prior to the establishment of Chilean industry, in the late 18th and early 19th centuries, the Juan Fernández Islands were already becoming con-nected to circuits of economic exploitation in the wider Pacific. Trade of the endemic sandalwood (*Santalum fernandezianum*) and the local fur seal (*Arctocephalus philippi*) pelts connected the islands to a transnational net-work of Pacific Island consumption, linking them with places such as the Marquesas, Hawai'i and the Fijian Islands. Alejandro Selkirk Island was transformed into a transoceanic sealing enclave starting in the 1790s, when sealers from the United States and Great Britain began to hunt the local fur seal in order to sell its desirable pelts in Eastern markets (Igler 2013, 115–117). The species was decimated to the extent that by the 1820s, it was almost extinct. While this population was able to recover, though, the exploitation of the Juan Fernández sandalwood was unfortunately more devastating. The tree began to be felled for its aromatic timber in the late 16th and early 17th centuries, and its exploitation peaked in the 1830s as it was commercialized in Asian markets (Johow 1896, 131; Venegas & Elórtegui 2022, 412). By the early 20th century, it no longer grew in the wild on the islands (Skottsberg 1918).

The arrival of tourism would build on both the existing connections to the Chilean mainland and to other Pacific islands, while also disrupting the traditional aquapelagic livelihood patterns that had waxed and waned over the previous century. Tourism had long been considered important by Chilean politicians and intellectuals. In his lengthy treatise on Juan Fernández, for example, influential Chilean statesman Benjamín Vicuña Mackenna popularized the idea of the islands as a romantic vacation spot (Vicuña Mackenna 1883). He believed that Juan Fernández's future was intimately linked with the development of tourist infrastructure, and Vicuña Mackenna's musings on the islands' "edenic" qualities would have a significant impact on how the archipelago was viewed in (and from) continental Chile. His views are echoed by journalist Jorge Guzmán Parada, who, in the 1950s, pondered the ideal conditions Robinson Crusoe Island provides for vacationing. In particular, he highlights how the proximity of the Juan Fernández archipelago to the continent and its everlasting spring weather makes tourism a potentially important source of income (Guzmán Parada 1951, 226). Although they are writing many decades apart, both Vicuña Mackenna and Guzmán Parada emphasize how tourism not only represents an economic opportunity for the development of the islands but is also instrumental for geoestrategical purposes. The sporadic fishing operations and penal colonies had meant that the islands were not securely settled, and tourism provided an opportunity to change this while also creating an additional income stream. Guzmán Parada, for example, compares their potential to the success achieved by the Hawaiian Islands as a tourist destination (1951, 226).

Especially in mid-century narratives, Juan Fernández's environment is depicted as an ideal setting for leisure: the island's temperate climate, impressive vistas and unique flora and fauna together provided the "most pleasant setting for the coexistence of animal, plant and human life" (Castedo 1949b, 53–54). Tourism represented the opportunity to capitalize on this unique nature, and to make it known to both global and domestic audiences, where, in the latter context, it would also facilitate the incorporation of Juan Fernández into the national geographical imaginary and the projection of Chile into the Pacific. The continental interest in marketing Juan Fernández offered the islanders a new commercial opportunity via the sale of local handicrafts and cuisine, which notably altered their aquapelagic engagements. While marine fauna had been the primary product of the extant colony, the arrival of tourism saw the increased commodification of terrestrial flora and fauna, most notably the Chonta palm and the Juan Fernández Firecrown Hummingbird (*Sephanoides fernandensis*), which became a coveted souvenir.

The first tourist cruise arrived at Juan Fernández on 5 January 1922. Aboard the Pacific Steam Navigation Company (PSNC)'s RMS *Ebro*, around 240 tourists were taken to the port of San Juan Bautista. While this was not a Chilean operation (the British PSNC being a subsidiary of the Royal Mail Steam Packet Company), the stop was added as part of the northbound leg PSNC's New York–Valparaíso route and featured many distinguished guests. The excursion attracted the cream of Chilean society, as Italian writer E. C. Branchi, who was on board, attests:

> Never has a steamer taken such a fantastic mixture of humans from the heart of the [Chilean] Republic as that which is travelling aboard our transpacific ship. Chile's "high life" has congregated on this deck.
> *(Branchi 1922, 126–127)*

In fact, the steamer was unable to accommodate all those who wished to visit the islands, and the PSNC sent the RMS *Essequibo* to repeat the visit from Valparaíso the following month (Unattributed 1922a, 11). The *South Pacific Mail*, one of the leading English-language newspapers in South America at the time, reported that passengers requested that the ship's captain sail around the islands before anchoring in port so that they might "admire its beauties and form a complete idea of the whole" (Unattributed 1922a, 11), a long detour which took from the morning until the late afternoon, reflecting the high levels of curiosity aboard. While Juan Fernández was not a regular stopover on the Valparaíso–New York route, it was added periodically to the itinerary and accompanied with special marketing that focused particularly on the archipelago's connection with Defoe's novel. In PSNC ephemera as in the English-language press, the islands are already referred to as "Robinson Crusoe's Islands" prior to the official renaming of Masatierra as Robinson Crusoe Island and Masafuera as Alejandro Selkirk Island by the Chilean Government in 1966. This difference in toponyms is just one example of the ways in which tourism has encouraged the Chilean state to "realign" the image of the islands with the image projected onto them, most notably by a British shipping enterprise.

Despite the success of the first tourist excursion, in the 1920s tourism in Chile was still a nascent industry hampered by a lack of infrastructure. This was compounded by the fact that it was relatively unknown as a tourist destination, both for domestic visitors and on the international market. After the first tourism law was passed in Chile in 1929, which promised to "[m]ake known, both within and outside of the Republic, the tourism centres and natural beauty of this country" (Biblioteca del Congreso Nacional de Chile

1929), the government sought to promote recently annexed and remote regions of Chile both to the metropolitan population and abroad (Booth 2010), producing bilingual English-Spanish promotional materials such as the photobook *Chile, país de belleza/Chile, country of beauty* (Servicos de Turismo 1937). Maritime transportation for tourists to Juan Fernández was more readily available from 1936, with both US-owned Grace Lines and the PSNC making the trip from Valparaíso. The fact that the islands became accessible via the likes of the luxurious PSNC MV *Reina del Pacífico* complete with cocktail lounge, library and ballroom, made them an ideal cruise destination both for short visits from the continent and as part of a longer South American itinerary, as such a trip could provide elements of "[f] antasy, escape, conspicuous extravagance and pampering" (Pirie 2011, 73). Long cruises were also an aspirational activity: drawing on the practices of 18th-century European aristocracy, partaking in a lengthy transatlantic or bicoastal American voyage was a way for businessmen to flaunt their wealth by taking up the leisure habits of high society (Williams 2003, 136). This was highly attractive to upper and middle-class Chileans as well, who saw partaking in tourism as directly relevant to class mobility and maintenance (Booth 2013, 134).

While Chile was not a well-known tourist destination in the early and mid-20th century, the Juan Fernández Islands' connection with the English literary canon superseded this issue. In Chile, British culture and capital had had a hegemonic influence since independence (Estrada 2006, 65). Given the profile of the novel, it is likely that many English speakers would have read or at least heard of Robinson Crusoe, and while Defoe situates his hero in the Caribbean, the possibility of seeing the place where Alexander Selkirk, upon whom the epic novel is based, had actually spent his four years and four months, gave the Juan Fernández Islands an appealing place in the domestic and international tourist markets. The role of privateering literature in shaping island imaginaries is not unique to Juan Fernández: Hayward and Kuwahara have identified a similar process unfolding in Takarajima in Japan, which is thought to have been the inspiration for the titular location of Robert Louis Stevenson's novel *Treasure Island* (1883) via an erroneous connection to the Scottish pirate Captain Kidd (c.1645–1701) (2014, 25). Notably, they highlight how even beyond the success of the novel in English and in translation in Japan, it became "productive for islanders to accommodate the association themselves" (Hayward & Kuwahara 2014, 28), thus reinforcing the island's connection to the exploits of a mariner who likely never set foot there and recalling the specious renaming of the individual Juan Fernández Islands in 1966. Alexander Selkirk actually spent his exile on what is now Robinson Crusoe Island, and not on the Island that today bears his name, Selkirk's literary incarnation having been deemed more important by the Government of Chile on account of its potential to attract tourists. Interestingly, though, it is a statue of Selkirk that stands in welcoming visitors today to San Juan Bautista (Robinson Crusoe Island) (Arana 2010, 295), documenting an embrace of this connection for the tourist market.

Chile had to market its desert and frigid extremes in creative ways in the early 20th century, creating the "Chilean Switzerland" in the southern region of Los Lagos, for example. In contrast, "Robinson Crusoe's Islands" were already alluring in the imaginations of many wealthy cruise goers keen to signal their familiarity with British and maritime cultural icons. This allure was not limited to the anglophone and anglophile markets, though. Since independence, many Chilean writers and statesmen have pondered the place of Juan Fernández as one of the nation's Pacific outposts. During the early republican period, the country's political projections were directed towards the continent. By the end of the 19th century, however, Chile's oceanic projections had become more important. After winning the War of the Pacific, fought against Peru and Bolivia (1879–1884), Chile became a regional power in the southeastern Pacific, further consolidating its power via the annexation of Rapa Nui (Easter) Island in 1888. Following this, Juan Fernández became an important part of the political discourse that defined Chile as a Pacific Ocean country and an insular nation. As they became incorporated into the national imaginary, the Juan Fernández Islands closely tied the country to Pacific island geographies, projecting Chilean power farther away from the South American continent.

Rising up from the sea as a volcanic island, the high levels of endemic species meant that the appearance of the Juan Fernández Islands was unsurprisingly not contiguous with that of continental Chile. The presence of the Juan Fernández sandalwood tree, in particular, revealed a possible connection with Asia and Australia, as it is not found anywhere on the American continent (Gálvez 1942, 59). Chilean writer Benjamín Subercaseaux, for example, highlights this aesthetic disjuncture, noting how the islands are "a land older than America" where one feels that "Chile is in its infancy" (1944, 200, 203), invoking their millennial past. They are "so near and yet so far" from Valparaíso, mused G. Mallett in the *South Pacific Mail* (1921, 13). Although cruise tourism was thus a key part of Chile's Pacific agenda, as Juan Fernández became a more popular and a more accessible destination, the distance between continental Chile and the islands would, antithetically, be further emphasized.

Fernándezian Relations: Lobsters, "Romance" and Cruise Ship Spatialities

While the livelihoods of the inhabitants of Robinson Crusoe Island, and indeed of the fictional Robinson Crusoe himself during his stint as a castaway, are directly tied to the sea, cruise ships actively reject engagement with the ocean, "attempting to remove and control the natural forces into which they come in contact" (Cashman 2013, 3). Writing predominantly of the contemporary mega ship, Cashman argues that the cruise ship is paradoxical in its ability to "form a unique type of aquapelagic assembly and to provide its antithesis" (Cashman 2013, 2). Although the *Ebro* and *Essequibo* predate the introduction of ship stabilizers in the 1930s, which helped passengers "forget they were in an aquatic environment" (Cashman 2013, 3), we can still see how early cruises facilitated non-engagement with oceanic and aquapelagic spaces. When visiting Juan Fernández, for example, tourists still slept in their cabins and mainly dined on board. It is further telling that the guests on board the first cruise in 1922 asked the captain to sail around the island so that they could view it. Instead of spending more time on the island itself interacting with the natural environment, they requested a view from the safe and comfortable setting of the deck. As Cashman continues "confinement and exclusion" defines life on board a cruise ship (Cashman 2013, 8), which is at odds with the lived experience of an aquapelagic society such as that of the inhabitants of Robinson Crusoe Island whose livelihood was directly tied to fishing.

The January 1922 visit marked the introduction of what is today one of Juan Fernández's primary industries. The arrival of cruise tourism also represents the conversion of the distinct spatialities of the aquapelago and the liner, as an episode from the first visit reported in the *New York Tribune* demonstrates. Titled "Robinson Crusoe Isle Lobsters, Big as Bull Pups, Attack Ship", the article describes events recounted by a Mr. Browne, who explains that upon arriving in Cumberland Bay (on Masatierra/Robinson Crusoe Island), the *Ebro* was swarmed by "lobsters as big as bulldogs climbing up the anchor chain and threatening the lives of every one aboard" (Unattributed 1922b).[1] The mayor, Alejandro McTush y Córdoba, "dressed in his bright seaweed suit and with a half coconut shell tipped over one eye", calls off the "attack", and explains to the passengers that the 200 or so inhabitants are waiting in the woods "to pray for the safety of the lobsters". The author goes on to explain how McTush y Córdoba demonstrated his method of luring the Juan Fernández rock lobster into lava pits as a cooking method (Unattributed 1922b, 20). While this tale conforms with stereotypical Euro-Western notions of unruly Pacific creatures that represent the border between "the known and the unknown and the natural and supernatural" (Byars & Broedel 2018, 3), the response of the islanders evidences a different kind of relationship that is representative of the "sacred, untouchable, memorised, and/or identified" elements that constitute an aquapelagic society as a space that is "generated by livelihood" (Suwa 2012, 15, 13). In this scene, the lobsters are revealed not as a passive presence in the constitution of the aquapelago, but as actants that impact upon the other constituent entities, transcending their role as sustenance and invading the tourist space. In line with Hayward's assertion that the concept of the aquapelago is uniquely positioned to help us understand humanity's changing relationship with the natural environment (2012b, 3), here we can see a distinct response from the visitors in contrast to the islanders' understanding of the lobster's role and agency in their conception of Fernándezian space.

The tourists are deeply unsettled by what they perceive to be an invasion of crustaceans that belong in the ocean onto the territory of the cruise ship – the *terra* root of this term being particularly key here. The lobster swarm disrupts the notion of the luxury liner as a space that replicates and is aligned with land-based amenities, such as ballrooms and smoking lounges, forcing the guests into a confrontation with the perceived threat of unfamiliar non-human life-forms that, in this account, extends to the islanders as well. The description of the inhabitants waiting in the woods evokes images of an Indigenous ambush, and it harks back to the European explorers killed on the shores of Pacific Islands such as James Cook and Ferdinand Magellan. Although many of these people would have been recent settlers from continental Chile or Europe, the article frames their interaction with the Fernándezian environment as naïve: they are "uncertain" about the impact of "a combination of steel and high class South Americans" on their primary food source, and have taken to "praying" for the safety of the lobsters (Unattributed 1922b, 20). This seems a reasonable response in the face of a massive, unfamiliar craft, the first large vessel to have visited the island in 42 years. What we can identify here, then, is the fashioning of the quaint yet threatening island space as in need of "protection from mainland interaction or contamination" (Baldacchino 2012, 57), which draws contingently on the "contrived exoticism" inherent in ocean cruising itineraries (Pirie 2011, 73). The islanders' relationship with the rock lobster represents a connection that the tourists perceive to be primitive and unnatural, while the crustacean transgression further disrupts the fallacious binary of land and sea creatures.

This connection of cruise tourism to European exploration of the Pacific is a key part of the "romance" that features heavily in tourist descriptions of Juan Fernández. A PSNC pamphlet advertising a 1937 "Sunshine Tour" around South America from Liverpool via the Strait of Magellan and Panama Canal onboard the *Reina del Pacifico* highlights that the archipelago is "rich in romantic associations" (7.2). This sentiment echoes longstanding descriptions of the islands' "romantic" appeal attested to by earlier Chilean authors such as Vicuña Mackenna, who quotes English Naval Officer George Shelvocke's 1726 description of Juan Fernández as "perfectly romantick" in his lengthy treatise on the archipelago (Vicuña Mackenna 1883, 153). In the PSNC pamphlet, this "romance" is connected explicitly to Defoe's work, and the whimsy of voluntary exile. By partaking in the cruise, visitors can follow in Selkirk's (and Crusoe's) footsteps, spending their own short period on the island before returning to the safety of the liner. Thus Selkirk's arduous journey from Europe is replicated in the leisure tourism context. Dean MacCannell has highlighted that a symptom of the transformation from industrial society to modern society is the transformation of work into a "touristic curiosity" (MacCannell 1999, 6): notably, visitors are following a trip Selkirk undertook as part of his employment as a coxswain. As such, Juan Fernández's enduring global connections, underpinned by labor, are refashioned from trade and privateering routes into cruise itineraries. The transoceanic journey required to reach the islands is no longer arduous and dangerous, but rather an experience to be indulged from the safety of a liner that strives to offer all the comforts of life on land.[2]

Crafting a Consumable Space: Juan Fernández and the Commodification of Endemic Species

The image of Juan Fernández propagated by the press and in tourist materials was not new. It reflected a scientific tradition that for decades had depicted the archipelago as a unique environment filled with exotic creatures, blessed with exquisite seafood and awash with ancient flora. Naturalists in particular constructed an image of the archipelago as an emblematic example of insular oceanic nature, connecting the islands' flora with the entire Pacific Basin and spotlighting links with places such as the Hawaiian Islands, New Zealand, Patagonia and even the Antarctic continent. In the 19th century, the Juan Fernández Islands attracted scientists from Europe and the Americas studying the natural history of Pacific islands and the ecology of insular environments. Expeditions such as the HMS *Challenger* voyage sponsored by the British Admiralty (1875), the Chilean botanical expedition led by Friedrich Johow (1891–1892) and the Swedish Pacific Expedition (1916) all remarked upon the unique natural history of the islands. They also praised Juan Fernández's natural beauty: Henry Moseley from the *Challenger* expedition was astonished by the islands' appearance, and the Swedish naturalist Carl Skottsberg went so far as to call Juan Fernández "the queen of an ocean" (Moseley 1879, 537; Skottsberg 1911, 148). Botanists and naturalists were drawn to the islands due to the high levels of endemism, particularly the vascular plant flora.

The construction of an idealized depiction of nature was not only a means of commercializing the islands for touristic purposes. The advent of cruise ship tourism also offered new opportunities for exploiting and commodifying the islands' natural elements. In this vein, discourses of Juan Fernández's nature fit well with Cheer, Cole, Reeves and Kato's characterization of the critical relationship between small islands and modern tourism: islands "have always maintained allure as sights of paradisiacal conceptualisations and the embodiments underlined by nature, remoteness and the 'island vibe' fit neatly into the touristic endeavour" (2017, 41). In the case of Juan Fernández, endemism was transformed into a selling point for tourists keen to have an "exotic" experience. TThe Chonta palm (Myrceugenia fernandeziana), the luma tree (*Myrceugenia schulzei*) and the local sandalwood were used to produce souvenirs that could be purchased by tourists. For example, islanders used Chonta wood and collected fossilized sandalwood to make walking sticks and aromatic boxes. They further cut down endemic ferns for tourists to use to decorate their homes (Ramírez 1935). Local animals were also sold as souvenirs, most notably the Juan Fernández Firecrown hummingbirds, which were desiccated by the islanders. They were popular with tourists on account of their attractive coloring and species dimorphism – red males and blue females – both with metallic "crown" plumage (Looser 1927, 243). *The South Pacific Mail* highlighted the local hummingbirds as a key attraction for nature lovers, remarking on the species' uniqueness (Unattributed 1921, 9). The tourist discourse thus transformed this bird species from an object of natural historical study into a static, commodified object of curiosity. Today, it is critically endangered.

Islanders profited financially from the new dynamics introduced by cruise ship tourism. By listening to the radio, the locals could find out when tourists would be arriving, and they would prepare the souvenirs in advance (Gálvez 1942, 60). In the press and in publicity, tourism was depicted as a natural extension of the islanders' hospitality. Guzmán Parada, for example, remarked how "all the inhabitants of the port crowd the boats, squeeze into the docks and climb into our little boat, to offer us their unreserved hospitality and cheerful cordiality" (Guzmán Parada 1951, 225). In turn, tourists were expected to buy these souvenirs, and, as travel accounts record, they gladly did so, bringing Chonta sticks, sandalwood memorabilia and even lobsters back to the ship (Branchi 1922, 124). In this project of transforming the island into a consumable and commodified space, elements of the underwater world were also marketable as goods, although they figured less predominantly than terrestrial items. Seafood became a key feature of tourist visits, where "[t]he lover of fish diet can secure his meal ashore, and he be regaled with rich luscious cod and lobster" (Unattributed 1921, 9). Significantly, seafood dishes would be many tourists' primary haptic and sensory engagement with the sea, representing one of their only forays into the islanders' aquapelagic society (Hayward 2012a; Suwa 2012).

As an idealized vision of nature, Juan Fernández was also advertised as a hub for botanists and amateur naturalists. Tourists participated in the epic narrative of how scientists such as Johow and later Skottsberg came across the last living sandalwood specimens. For decades, naturalists from Europe and the Americas had struggled to understand why it started to decline in population in the early 19th century. Obsessed with this question, Johow conducted several expeditions to the islands in the 1890s, and he and his crew were able to find one of the last living specimens on a Fernándezian peak. By analyzing this tree, Johow concluded that the local sandalwood had become extinct due to intensive human exploitation as it was felled and sold to international markets (Johow 1896, 130). The epic tale of how Johow was able to unravel this mystery was regaled to visitors by Pedro Arredondo, a local islander who guided the botanist through the lush forests and steep summits (Gálvez 1942, 59). Yet this story was not framed as a cautionary tale about human depletion and overconsumption; rather, its aim was to entertain tourists and underscore the idea of naturalists as modern heroic figures who struggle against nature in their quest to understand it. Furthermore, narratives such as Johow and Arredondo's tended to overemphasize the local connections with the land, invisibilizing the importance of the waterscape for their livelihood and everyday activities.

There were, however, early critics of the impact of tourism on the Juan Fernández Islands, particularly among the scientific community, which was was concerned about biodiversity loss and the overconsumption of endemic species. Chilean naturalist Gualterio Looser, for example, underscored how intensive tourism had had a considerable impact on the islands' environment. Even though he had experienced cruise ship tourism to the island himself, embarking three times between 1925 and 1927, Looser noted in 1927 how in the two years since his first visit, the Chonta palm had all but disappeared. He opined that the Chonta was in grave danger and feared that the continued exploitation would result in its disappearance (Looser 1927, 242). Similar critiques were levelled by the Chilean biologist Filomena Ramírez. Writing in 1935, Ramírez argued that if the exploitation of the local palm was maintained at its current rate, the Chonta would inevitably meet the same fate as the sandalwood (1935, 58). Ramírez, though, went further in her criticism. At the April 1935 meeting of the Chilean Society of Natural History, Ramírez presented an article that detailed the extinction risk of the Chonta palm, advocating for the production of improved knowledge about the islands' environment and the need to implement conservation policies. As Looser had previously noted, she contested that tourism accentuated the overexploitation and overconsumption of certain endemic species. Today, many of the flora exploited for tourist consumption, including the Chonta and the Luma, remain extremely vulnerable.

The designation of the Juan Fernández Islands as a National Park in 1935 did not have an impact on the social and cultural representations of the islands or their economic activities for at least three decades. Cruise ships continued to stop in Cumberland Bay without regulation. The decision to transform these islands into a protected area was entangled with geopolitical aims, economic development projects, including the promotion of tourism, and efforts to preserve endemic nature. As Magdalena García and Monica

Mulrennan's recent research has demonstrated, the Chilean state developed its first national parks "to assert political sovereignty over remote territories and resources" (García & Mulrennan 2020, 199). Environmental protection also offers new ways of controlling people's activities and behaviors. Baldacchino has previously connected the creation of island imaginaries to political control (2012, 57). In the case of Juan Fernández, the creation of the national park provided the opportunity to protect the forests and control the economic production of the islanders. Indeed, the 1935 decree that established the islands as a national park specifically prohibited the exploitation of the Chonta palm and endemic ferns, recognizing the unique nature of the Islands (Ministerio de Tierras y Colonización 1935). Yet this protection was not enforced: Juan Fernández's nature continued to be exploited in order to create souvenirs into the mid-20th century, as Guzmán Parada's account attests (1951, 226). As Federico Freitas has acknowledged, "establishing protected areas only on article made sense, as national governments could reap the benefits of a token of 20th-century-modernity – the national park – without incurring the political and economic costs of implementing it" (Freitas 2021, 7). All things considered, the creation of Juan Fernández Archipelago National Park offered the opportunity to reinforce Chilean sovereignty over the islands, bringing them "closer" to the national imaginary and integrating them as one of a series of idiosyncratic natural environments, such as the Patagonian forests and the tropical islandscape of Rapa Nui. Remoteness, then, was not perceived as hindrance to national unification, but as an asset for projecting Chile farther into the Pacific that would contingently advance the national tourism agenda.

Juan Fernández has long been at the forefront of conservation efforts both in Latin America and beyond. Efforts were made to protect the archipelago's marine resources as early as 1907. This constituted a Chilean Government moratorium on lobster fishing from September to December each year during the breeding season and regulation of the size of the lobsters that could be caught (Arana 2010, 226). This has been a successful policy that remains in place today and is actively supported by the islanders. Further protections have been implemented as the development of marine biology and other ocean sciences nationally and internationally over the course of the 20th century has enhanced knowledge of the rich and biodiverse underwater world surrounding the archipelago. Scientists have identified a unique mix of tropical, subtropical and temperate species of flora and fauna, including many endemic species such as the Juan Fernández fur seal (*Arctocephalus philippi*), Chile's only endemic pinniped. In 1977 UNESCO granted the archipelago the status of World Biosphere Reserve, elevating the global importance of this unique ecotonal ecosystem.

This status has not been sufficient to protect the archipelago from over-exploitation, particularly overfishing. In response to decades of decline in the seabass population, local fishermen stopped fishing for commercial purposes, although they continue to use seabass as bait for trapping lobsters and for their own consumption (Arana 2010, 262). Despite the islanders' and the government's work to end overfishing, in recent decades industrial deep sea fishing of the Orange roughy (*Hoplostethus atlanticus*) and the Splendid alfonsino (*Beryx splendens*) in the unprotected surrounding waters has depleted the marine environment further. A recent study by Porobic et al. (2019) has since identified the Juan Fernández Islands as a Vulnerable Marine Ecosystem (VME), highlighting that this environment is easily disturbed and has a low rate of recovery. In 2018 the Government of Chile implemented a world-leading policy that saw the creation of a marine park around the islands to prevent further damage, encompassing a $262,000m^2$ "no take zone" where commercial fishing is prohibited, but artisanal fishing can continue. This turn to maritime conservation is a recognition of the ecotonal nature of this environment, counteracting the terrestrial bias of previous conservation policies.

This proposal emerged from the efforts of local islanders' organizations: the Sindicato de Pescadores Independientes de la Pesca Artesanal de Juan Fernández [Juan Fernández Artisanal Fishing Independent Fishermen's Syndicate], the Agrupación de Pescadores de Robinson Crusoe [Robison Crusoe Fishermen's Association], the Sindicato de Pescadores de la isla Alejandro Selkirk [Alejandro Selkirk Island Fishermen's Syndicate], the Fundación Endémica [Endemic Foundation], and the Gremio de Turismo de Robinson Crusoe [Robinson Crusoe Tourism Union], as well as the support of international environmental organizations, including Oceana, National Geographic Pristine Seas, Island Conservation, Oikonos and the Government of Chile. This bottom-up initiative, which preserves local fishing practices while protecting the area from overfishing, has transformed Chile and the Juan Fernández Islands into global leaders in marine conservation, at least on paper. Although the inclusion of Juan Fernández in a Marine Protected Area (MPA) offers it legal protection, and lobster and fish populations remain stable, Southworth (2021) has emphasized that this MPA is "designated but unimplemented", meaning that extractive practices unfortunately do still occur.

Conclusion

Returning to Macleod's notion of "cultural realignment", this chapter has shown how the arrival of cruise tourism in Juan Fernández led to the refashioning of its identity in hegemonic cultural contexts as a remote, edenic outpost untainted by human influence. In spite of the predominant image of isolation, the creation of Juan Fernández as a quasi-tropical cruise destination inserted the islands into new global circuits of travel that built on the successful transatlantic and transpacific shipping routes developed in the 19th century. The cruise liner in particular, as we have demonstrated, implies a negotiation of aquapelagic space that can make tourists feel uncomfortable as what they perceive to be stable boundaries between land and sea give way, especially in the disquieting liminal space of the shore, which as Gillis has emphasized has often been a site of transformation, terror and the supernatural (2012). As such, we can see how the visions of Juan Fernández in travel accounts and tourism marketing fail to recognize the traditional aquapelagic orientation of the island. While the tourists understand the water-land boundary as static, the islanders' multifarious relationship with the rock lobster proves that this is not the case. In the face of tourism, though, their aquapelagic relationality is reconfigured as new opportunities for commerce encourage further commodification.

Cruise tourism discourses reproduced colonial imaginaries that focused on land-centric depictions of the islands' nature. These discourses engaged with triumphalist representations of the islands' past and created romanticized depictions of its unique nature that focused on its connection to an English literary hero. The Government of Chile's decision to rename Masafuera as Alejandro Selkirk Island, a place where he never set foot, in order to reserve "Robinson Crusoe" for the primary tourist destination of Masatierra, where he actually resided, betrays the perceived importance of this historical connection and its ability to reach across bodies of water. The Juan Fernández Islands' historical relationship with the Chilean state is thus complex, particularly as the nation has long struggled to negotiate its maritime image. Chile has constructed its identity with its back to the Pacific Ocean; the country's oceanic character and insularity did not start to figure in the national geographical imaginary until long after independence (Peliowski & Valdés 2014). Yet, as Chile became a regional power following its victory in the War of the Pacific, national discourses started to emphasize the importance of its Pacific positionality. Juan Fernández thus became an important player in the national agenda. In this context, tourist practices were instrumental to the country's maritime claims, giving more importance to oceanic islands such as Juan Fernández and Rapa Nui, which had long been considered remote, marginal and unimportant for the country's development. Chilean Pacific islands began to be branded as exotic and romantic destinations, a taste of paradise that represented the interface between the national territory and maritory.

Finally, the overexploitation and overconsumption of species endemic to Juan Fernández, such as the Firecrown and the Chonta palm tree, can be seen as cautionary tales documenting the impact of modern tourist practices on small islands. While each visit only brought a few hundred travellers, early-20th-century cruise tourism had a remarkable impact on the island environment. Early criticism of tourism levelled by naturalists such as Looser

(1927) and Ramírez (1935) showcase the rapid declines caused by uncontrolled commodification. As tourist activities are still of economic and social importance in the Juan Fernández archipelago, it seems pertinent to close by reflecting on the enduring cultural and environmental impact of tourism. Thankfully, conservation policies covering both marine and terrestrial environments implemented by the National Park administration in recent decades, and active efforts to create sustainable economic practices, particularly in lobster fishing, have helped to preserve the island's endemic species. Unlike in the first half of the 20th century, current environmental and social discourses are seeking to understand the island as a fluid, continuous and porous space at the juncture of land and sea where human and non-human agents interact. The efforts in Juan Fernández represent a key example of how a local island population can advance a progressive conservation agenda. While there is still work to be done, enforcing the Marine Protected Area will ensure that the islanders' aquapelagic relationality can continue for many years to come.

Declaration

This article is part of the postdoctoral research project N° 3220196 funded by the Chilean National Agency of Research and Development (ANID).

Notes

1 This article was reprinted in the *South Pacific Mail* on 16 March 1922, p. 18.
2 Please see our article "Reimagining the Juan Fernández Islands: Cruise Tourism and the Commodification of Nature", *Shima* 17(1), 6–22, for an in-depth analysis of promotional materials, including the PSNC pamphlet.

References

Arana, Patricio, M. 2010. *La Isla de Robinson Crusoe*. Ediciones Universitarias de Valparaíso.
Baldacchino, Godfrey. 2012. The lure of the island: A spatial analysis of power relations. *Journal of Marine and Island Cultures*, 1(2), 55–62.
Biblioteca del Congreso Nacional de Chile. 1929, February 13. *Ley 4585*. https://www.bcn.cl/leychile
Biblioteca del Congreso Nacional de Chile. 2020, June. *Reportes estadísticos 2021 de Juan Fernández*. https://www.bcn.cl/siit/reportescomunales/comunas_v.html?anno=2021&idcom=5104
Booth, Rodrigo. 2010. El paisaje aquí tiene un encanto fresco y poético': Las bellezas del sur de Chile y la construcción de la nación turística. *HIB: Revista de Historia Iberoamericana*, 3(1), 10–32.
Booth, Rodrigo. 2013. More than an exotic place to visit: Recent contributions to the history of tourism in the Southern Cone (Argentina, Chile and Uruguay). *Mobility in History*, 4(1), 129–135.
Bremner, Lindsay. 2017. Observations on the concept of the aquapelago occasioned by researching the Maldives. *Shima*, 11(1), 18–29.

Burke, A. 2020. An aquapelagic evolution? Developing sustainable Tourism Futures in Galápagos, Ecuador'. Shima 14 (2), 57-80.

Byars, Jane. and Broedel, Hans Peter. 2018. Introduction. In *Monsters and borders in the Early Modern imagination*, 1–18.

Camillo, Branchi, E. 1922. *La Isla de Robinson. Impresiones de arte y de vida de un extranjero en Chile*. La Patria.

Carmona Jiménez, J. C. 2021. Las islas en un museo del continente: Colonialidad e imágenes del archipiélago de Juan Fernández. *Revista Márgenes: Espacio, Arte y Sociedad*, 14(1 20), 57–67.

Cashman, D. 2013. Skimming the surface: Dislocated cruise liners and aquatic spaces. *Shima*, 7(2), 1–12.

Castedo, L. 1949a, March. La naturaleza y el hombre en Juan Fernández. *En Viaje*, *185*, 54–55.

Castedo, L. 1949b, May. La naturaleza y el hombre en Juan Fernández. *En Viaje*, *187*, 53–54.

Cheer, Joseph. M., Reeves, Keir. J., Cole, Stroma., & Kato, Kumi. 2017. Tourism and islandscapes: Cultural realignment, social-ecological resilience and change. *Shima*, *11*(1), 40–41.

Freitas, Frederico. 2021. *Nationalizing Nature: Iguazu Falls and National Parks at the Brazil-Argentina border*. Cambridge University Press

Gálvez, O. E. 1942, September. Una excursión a Juan Fernández. *En Viaje, 107*, 58–61.

García, Magdalena. and Mulrennan, Monica., E. 2020. Tracking the history of protected areas in Chile: Territorialization strategies and shifting state rationalities. *Journal of Latin American Geography*, 19(4), 199–234.

Gillis, J R. 2012. *The human shore*. University of Chicago Press.

Gillis, J R. 2014. Not continents in miniature: Islands as ecotones. *Island Studies Journal, 9*(1), 155–166.

Guzmán Parada, J. 1951. *Cumbres oceánicas. Las Islas Juan Fernández*. Bustos y Letelier.

Hau'ofa, Epeli. 1994. Our sea of Islands. *The Contemporary Pacific, 66*(1), 1–11.

Hayward, Philip. 2012a. Aquapelagos and aquaplegic assemblages. *Shima, 6*(1), 1–14.

Hayward, Philip. 2012b. The constitution of assemblages and the aquapelagality of Haida Gwaii. *Shima, 6*(2), 1–14.

Hayward, Philip and Kuwahara Sueo. 2014. Takarajima: A treasured island: Exogeneity, folkloric identity and local branding. *Journal of Marine and Island Cultures, 3*, 20–30.

Igler, David. 2013. *The Great Ocean: Pacific worlds from Captain Cook to the Gold Rush*. Oxford University Press.

Johow, Friedrich. 1896. *Estudios sobre la flora de Juan Fernández*. Imprenta Cervantes.

Looser, G. 1927. Sobre algunos objetos que venden los habitantes de las islas de Juan Fernández—Apuntes folklóricos. *Revista Chilena de Historia Natural*, 31, 240–244.

MacCannell, Dean. 1999. *The tourist: A new theory of the leisure class*. University of California Press.

MacLeod, Dean. 2013. Cultural realignment, Islands and the influence of tourism: A new conceptual approach. *Shima*, 7(2), 74–91.

Mallett, G. 1921, September 29. Juan Fernández: The latest chapter in the history of Robinson Crusoe's Island. *South Pacific Mail*, 13–14.

Ministerio de Tierras y Colonización. 1935. *Decreto 103*. 16 January 1935. https://www.bcn.cl/leychile/navegar?i=269837.

Moseley, Henry Nottage. 1879. *Notes by a naturalist on the 'Challenger', being an account of various observations made during the voyage of the H.M.S. Challenger round the world, in the years 1872–1876.* Macmillan and Company.

Pacific Steam Navigation Company. 1936. *Ideal sunshine tour round South America, Robinson Crusoe's Island, & West Indies.* SAS/33F/4/4 - Merseyside Maritime Museum, Liverpool.

Peliowski, Amari Vald and Valdés, Catalina, eds. 2014. *Una geografía imaginada. Diez ensayos sobre arte y naturaleza.* Universidad Alberto Hurtado-Metales Pesados.

Pinset, Guillermo, B. 2007. De colonos a endémicos: La identidad naturalizada en la Isla Robinson Crusoe, Archipiélago Juan Fernández. *VI Congreso Chileno de Antropología. Colegio de Antropólogos de Chile A. G*, 1733–1740. http://bibliotecadigital.academia.cl/xmlui/handle/123456789/6195

Pirie, Gordon H. 2011. Elite exoticism: Sea-rail cruise tourism to South Africa, 1926–1939. *African Historical Review*, 43(1), 73–99.

Porobic, Javier, Fulton, Elizabeth, Parada, Caroline et al. 2019. The impact of fishing on a highly vulnerable ecosystem, the case of Juan Fernández Ridge Ecosystem. *PLoS One, 14*(2), 1–32.

Ramírez, F. 1935. Mi último viaje a Juan Fernández. *Revista Chilena de Historia Natural, 39*, 57–59.

Robinson Wright, Marie. 1904. *The Republic of Chile: The growth, resources, and industrial conditions of a great nation.* G. Barrie & Sons.

Scribner, Vaughn. 2021. From sailor traps to tourist traps: Mermaid-themed tourism destinations in the United States of America. *Shima*, 15(2), 72–83.

Servicios del Turismo. 1937. Chile, *país de belleza/Chile, country of beauty.* Instituto Geográfico de Agostini.

Skottsberg, Carl. 1911. *The wilds of Patagonia: A narrative of the Swedish expedition to Patagonia, Tierra del Fuego and the Falkland Islands in 1907–1909.* Edward Arnold.

Skottsberg, Carl. 1918. The Islands of Juan Fernandez. *Geographical Review*, 5(5), 362–383.

Southworth, Grace. 2021. Marine protection of the Juan Fernández Islands. *ArcGIS StoryMaps*, 25 May 2021. https://storymaps.arcgis.com/stories/8500ba0b1404463f8cb44153c5234691

Stevenson, R.L. 1719. *The Life and Strange Surprizing Adventures of Robinson Crusoe, of York, Mariner.* William Taylor.

Subercaseaux, Benjamin. 1944. *Chile, o, una loca geografía* (6th ed.). Ediciones Ercilla.

Suwa, Jun'ichiro. 2012. Shima and aquapelagic assemblages. *Shima: The International Journal of Research into Island Cultures*, 6(1), 12–16.

Turra, Baldomero, E 2006. La colectividad británica en Valparaíso durante la primera mitad del siglo XX. *Historia*, 1(39), 65–91.

Unattributed. 1921, December. Westward Ho! Some notes on Crusoe's isle. *The South Pacific Mail*, 9.

Unattributed. 1922a, January. The pleasure cruise of the 'Ebro' To Robinson Crusoe's Islands. *South Pacific Mail* (accessed via Readex Latin American newspapers).

Unattributed. 1922b, February 2. Robinson Crusoe Isle lobsters, big as bull pups, attack ship. *New York Tribune*, 6.

Venegas, Fernando V. and Elórtegui, Sergio. 2022. La Huella Humana en la Isla Robinson Crusoe (Archipiélago de Juan Fernández) entre los Siglos XVI y los Albores del Siglo XVII: Una Impronta para el Futuro' Historia Ambiental Latinoamericana y Caribeña (HALAC) revista de la Solcha. 12(1). 388–43

Vicuña Mackenna, Benjamin. 1883. *Juan Fernandez, historia verdadera de la isla de Robinson Crusoe*. R. Jover.

Williams, David M. 2003. The extent of transport services' integration: SS Ceylon and the first "Round the World" Cruise, 1881–1882. *International Journal of Maritime History*, 15(2), 135–146.

Woodward, Ralph Lee. 1969. *Robinson Crusoe's Island. A History of the Juan Fernández Islands*. The University of North Carolina Press.

8

THE ENTANGLED ISLAND

Katchatheevu and Indo-Lankan Maritime Relations

Arup Chatterjee

Prologue

Approximately 20 miles north of the tip of the Dhanushkodi Peninsula, in India's state of Tamil Nadu, and 20 miles to the southwest of Sri Lanka's Delft Island, lies a contested island known as Katchatheevu (which translates to "barren island" in Tamil) (Fig. 8.1). It is a 285-acre uninhabited isle formed by a volcanic eruption in the 14th century. Shrouded in obscurity, Katchatheevu has a knack for resurfacing, such as in 2023, when Indian prime minister Narendra Modi indirectly referenced it in a speech in the Indian lower house of parliament, the Lok Sabha.

Katchavtheevu is a "square-shaped island" that is "one-fifth as large as New York City's Central Park … one-half mile long and barely one-half mile wide" ("Crisis over 150 acres", 24). During the colonial era it was used as a military base by the British army and is now the location of a small Catholic Church (3.6 by 4.3 meters) devoted to St. Anthony, the patron saint of fishermen and travellers. Built in 1905 by Tamil merchant Seenikuppan Padayachi, the church was intended as a retreat for Indo-Ceylonese fisherfolk caught in turbulent weather or in need of drying fishing lines, and it is visited by Tamil and Lankan pilgrims every year, at the end of March, for a week-long religious festival. The island is believed to have served as a smugglers' base until its geopolitical dividends became amplified in February 1968 when Indian prime minister Indira Gandhi ceded about 350 square miles of arid territory in the Rann of Kutch region in Gujrat to Pakistan (Phadnis 1968, 783, 788; Anand 1968). Sensing a parallel prospect in the Gulf of Mannar, Ceylon began laying claims over Katchatheevu on account of St. Antony's Church's affiliation with the northern Jaffna's Roman Catholic

DOI: 10.4324/9781003569534-8

FIGURE 8.1 Map of Kachatheevu's position between Sri Lanka and southeastern India. (Google Maps)

diocese. Although Katchatheevu was essentially a moorland of cacti without drinking water and too insignificant to be seen on most postcolonial maps, Lankan demands for the island eclipsed the then escalating problem of stateless Tamil refugees that confronted both India and Sri Lanka.

Eventually, in 1974, Katchatheevu was conceded to Sri Lanka, by the Indira Gandhi–led Indian administration, in furtherance of the mutual goodwill she shared with the then-Lankan prime minister, Sirimavo Bandaranaike. The cession arose out of the Indo-Sri Lankan Maritime Agreement that promised to resolve discords in the Palk Bay region. In India, it was celebrated as a riposte to "the canard that India behaves overbearingly towards its small neighbours", and, even though the Indian administration believed that it possessed "an unassailable case", it claimed to have forfeited Katchatheevu in order to obtain "harmonious relations with Sri Lanka" (*Indian and Foreign Review*, 24). The cession preceded the 1976 exchange of letters between the two nation-states that delineated the maritime boundary line in the Sethusamudram littoral region. Since the onset of the Lankan civil war in 1983, the island became an unofficial theater for clashes between Indian Tamil fishermen and a predominantly Sinhalese Lankan navy. These conflicts have resulted in the loss of livelihoods, properties and lives of Indian fishers due to unintended forages across the international maritime boundary. Recently, Sinhalese fishermen have voiced concerns that the Lankan administration could be politically persuaded to lease the island to India as a compensatory measure. However, the Katchatheevu situation is far too complex to be reduced to nationalistic rhetoric or parochial concerns. It is a relic of the complex legacy of geopolitical challenges stemming from colonial South Asia. Besides, in being a complex of human and non-human island disputes, it also exemplifies the notion of the aquapelago.

The term "aquapelago" connotes assemblages of human relations with marine and land spaces in and around islands and littoral waters. Along with being a perspective on the destinies of sociopolitical lives set around "waters between and waters encircling and connecting islands", the aqua-pelago operates as "a social unit" wherein "aquatic spaces between and around a group of islands are utilized and navigated in a manner that is fundamentally interconnected with and essential to the social group's habi-tation of land and their senses of identity and belonging" (Hayward 2012a, 5; Hayward 2012b). Informed by this definition, this chapter offers the his-tory of Katchatheevu to argue that official – and probably even discursive – political lexicons of communicating about islands and islandic identities can blur their deeper spatiotemporal entanglements with other geographies and histories, especially the geo-historicity of human–non-human relations in precarious spaces of the Anthropocene.

The Lankan navy's antipathy towards Indian fishers' rights, alongside continuous inadvertent border crossings by both Indian and Lankan fishers across the International Maritime Boundary Line, constitute glaring evi-dence of the mutual derecognition of the geopolitical barriers in the psyche of Tamil (and even Sinhalese) fishers on both sides of the border. The 1974 and 1976 agreements were unsuccessful in specifying "the fishing rights of the fishermen of Tamil Nadu" (Joshi 2022), which led to Katchatheevu becoming the fulcrum of transgressive mobilities of Tamil fishers. Rather, the diurnal rhythms of their littoral life may also be said to be transgressed by the official laws of the two postcolonial nation-states (Stephen 2015, 71). Meanwhile, India and Sri Lanka have resumed strong political ties since about 2009 (culminating in India's unconditional economic aid to its littoral neighbor in 2021–2022). Nevertheless, the impact of the nation-state–level bilateral trade and traffic is at variance with the domestic and civil society perceptions of Indian and Lankan political attitudes regarding each other. In the process, Katchatheevu has become an Achilles' heel of nationalist discourses from both postcolonial nations. Posing the question of who pos-sesses Katchatheevu ignores the aquapelagic underpinnings of the island – its potential to be simultaneously performed as a geopolitical territory in bilat-eral political and nationalist rhetoric while it is precariously experienced as a shared maritime geoheritage of Tamil legacies across the Indo-Lankan international maritime boundary line.

Entangled Conflicts

India's cession of Katchatheevu was (as claimed by the then Indian administration) an act of bilateral goodwill. The transaction went ahead without Indian parliamentary proceedings, which, eventually, aggravated the challenges of Pamban's fisherfolks and other native stakeholders. Tamil fishers' memories of Katchatheevu's loss have never really subsided. They reappear every time discussions of the controversial Sethusamudram Shipping Canal project are resumed in the Tamil Nadu legislative assembly. The Sethusamudram project envisages a continuous navigable marine passage between India and Sri Lanka, across the Palk Strait. Its idea dates back to James Rennel's surveys of Adam's Bridge in the 1770s. The project remains an old dream of Tamil political leaders. The Sethusamudram Project was stalled in 2007–2008 by judgments of the Supreme Court of India, and in subsequent judgments by the apex court, on grounds of religious and ecological sensitivity.

Many Indians, predominantly of the Hindu faith, believe that the narrow strip of the submerged tombolo known as Adam's Bridge (Chatterjee 2022, 2024), connecting India and Sri Lanka, harks back to a legendary bridge believed to have been constructed in primitive times by Lord Ram and his hominoid army, according to the Indian epic *Ramayan* authored by sage Valmiki. Abrahamic traditions associate the structure with the legend of Adam being expelled from Paradise, following which he descended on Adam's Peak or Sri Pada, in Sri Lanka, and crossed the Palk Strait over to India via a primitive "bridge" across the seabed (Field 1903, 39–40; Paranavitana 1958; 47; Dunnett 2021). Contested mythologies – albeit by no means insubstantial in Sethusamudram's history – constitute only the tip of the iceberg, which, in this case, comprises an interspecies conflict in the Palk Bay region. The region in and around the Palk Strait is especially ecologically vulnerable and historically prone to earthquakes and tsunamis. Allied to this is the mammoth challenge of the sinking islands of the Gulf of Mannar, owing to Anthropocene coral deforestation, partly due to coral and sea-wealth exploitation by fisherfolk and sea-facing communities – a problem exacerbated by scarcity of resources due to monopolies over the Sethusamudram region by large corporations, businesses and trawling operations. This trialectic of faith, ecology and political economy inevitably structures Katchatheevu's destiny, while Tamil Indigenes living in close proximity and dependent on the aquapelagic assemblage of the Sethusamudram region for their subsistence and cultural memories have become distant others in the discourses around the contested island.

For geopolitical observers of islands, Katchatheevu is liable to educe other island conflicts of recent times – with which the Sethusamudram region seems entangled. The reverberations of the Beagle conflict between

Argentina and Chile – settled in 1984 through a papally mediated Treaty of Peace and Friendship after a prolonged dispute over the Picton, Nueva and Lennox Islands – are still felt in contemporary South American politics (Van Aert 2016). Nearer Katchatheevu lie the uninhabited and contested Senkaku Islands, situated in the East China Sea. Known as the Diaoyu Islands in China and as the Tiaoyutai Islands in Taiwan, they have been administered by Japan since the 19th century, from the time of the archipelago's Japanese possession, owing to it being *terra nullius*. However, besides, being ruled by the United States between 1945 and 1972, the islands have been subject to disputed claims by China and Taiwan since the 1970s, leading to observers suggesting that the archipelago should be considered amenable to heterogenous truth claims and shared jurisdiction (Ong & Perono Cacciafoco 2022), an aspect that governs its entanglement with other conflicted islands, including Katchatheevu. Likewise, Pedra Branca, a Singaporean island, presents a somewhat similar, counterintuitive situation. In 1979, the island became the nucleus of a bilateral dispute when a map of Malaysia's territorial waters and continental shelf boundaries depicted it as part of Malaysian territorial waters. Malaysia justified it by virtue of its historic claim over the state of Johor. Nevertheless, in May 2008, the International Court of Justice ruled in favor of Singapore's claim to Pedra Branca. All the same, both states agreed to allow shared fishing rights in and around Pedra Branca without imposing restrictions on each other. In 2017, Malaysia brought the case back to the Court for a reinterpretation of the judgment, only to withdraw its applications in 2018. The withdrawal was explained by the Malaysian prime minister as an exemplary demonstration of regional cooperation and respect. In the course of the focus on Pedra Branca, its precolonial history came to light, shifting attention from geostrategy to its paleogeographic and maritime history. This included relatively abject themes like shipwrecks, including those of medieval Chinese vessels (Baldacchino 2016) – a fact that underscores other Southeast Asian legacies of the island.

These loosely bonded themes of littoral conflicts and their geo-historical milieus have more than a passing impact on Katchatheevu's discursive destiny. These outlier island stories, otherwise "deemed unworthy of the professional scholar's attention", represent a Foucauldian "insurrection of subjugated knowledges" or of "imaginations that have been discarded or disavowed by the disenchanted disciplines within which we function as professional scholars" – as Sumathi Ramaswamy stated in her history of the fabulous sunken continent of Lemuria (Ramaswamy 2004, 4). Concurrently, the story of Katchatheevu also requires one to, as Prathama Banerjee has suggested elsewhere, "embark on an adventurous time travel that makes us not just free political subjects but also free theoretical subjects" (Banerjee 2020, 220) – within a subjecthood that does not make Katchatheevu an extractible field for idioms, ideologies or elements of political interest to

21st-century nation-states. Besides the fact of its controversial cession, in the last 50 years, or so, Katchatheevu's status has been politically amplified by the United Nations Convention on the Law of the Sea (1982–1994) – a progeny of the third United Nations Conference on the Law of the Sea (1973–1982) and the BBNJ High Seas Treaty (United Nations 2023) – that pursues regulations and legislation for environmental and geopolitical issues surrounding international marine territories. Katchatheevu finds itself embroiled in the metamorphoses of borderland islandscapes into zones of ethnic, economic or armed conflicts, intensified by pro-self-rule rhetoric. In the wake of decolonization, the colonial imagination of boundaries demarcating the territorial extents of India and Sri Lanka was deployed in the nationalistic rhetoric of both nations, even as the newly ordained territories kept on being radically tested by the littoral forays of Indo-Lankan fishers. These littoral negotiations – while unrecognized in the official postcolonial discourses of the two nation-states – pose a dynamic reassessment of questions of the ownership of Katchatheevu. The island's extant geopolitical limits are incapable of eradicating its ethno-cultural and ecological ambivalence (Sherif 2013). Indo-Lankan Tamil fishers "fall under the category of contiguous borderland people who have established linkages since colonial and pre-colonial times" (Sherif 147). Their cross-cultural navigations elicit "new regional, continental, and transcontinental routes of connection" that intermesh Indo-Lankan waters, territories, fish, geologies and cultures in an "uncanny stretching and overlapping of geographies" (Mezzadra & Neilson 2013, 212).

The Colonial Legacy

Katchatheevu was historically a territory constituting 69 seaside villages and 11 islets of the Ramnathapuram (Ramnad) Zamindari (a semi-autonomous feudal estate). The *zamindari* was instituted in 1605 by Madurai's Nayak dynasty, and it was owned by the Sethupathis – a title literally meaning "the protectors of the Sethu" (Adam's Bridge). The Sethupathi ruler, Koothan Sethupathi – who ruled Ramnad between 1622 and 1635 – is known to have commissioned a copper tablet whose inscription confirms that the jurisdiction of the territory extended to Thalaimannar in Sri Lanka. Katchatheevu returned substantial economic yields for its Sethupathi rulers. The Dutch East India Company leased the island, in 1767, from the then Sethupathi ruler, Muthuramalinga Sethupathi. Subsequently, in 1822, the British East India Company rented it from Ramaswami Sethupathi. In 1845, three proclamations by the Governor of Ceylon, Colin Campbell, defining the boundaries of Jaffnapatnam, omitted Katchatheevu, since the island was not considered a part of British Ceylon (Jayasinghe 2003, 74). In July 1880, Muthuswamy Pillai and Muhammad Abdul Kader Maraickar of the Madras Presidency signed a registered lease deed for five years granting Edward Turner, the Special Assistant Collector of Ramnad (under the Zamin's Court of Wards). The lease permitted root collections, for manufacturing dyes, from Ramnad's 70 villages and 11 islands, including Katchatheevu. Similarly, in 1885, the merchant Ramaswamy Pillai drew up an agreement for a comparable venture and duration in favor of the manager of the Ramnad estate, T. Rajarama Rayar. It also included Katchatheevu as part of the lease. In 1913, a lease executed between the Ramnad king and the Secretary of State for India granted exploitation rights over chank shells within the specified limits, explicitly recognizing Katchatheevu as being located on the Indian side of the Palk Bay. This reaffirmation of its status as being part of India rather than Ceylon was not without its precedents. The prevailing commonsense was solidified by the judgment in the Annakumaru Pillai v. Muthupayal case (1904). The Madras High Court's verdict in this case denoted that Katchatheevu was "an integral part of His Majesty's dominions" while its adjoining chank beds "constituted the territories of British India" (Suryanarayan 2005, 69). This decision, and a 1922 report from the Imperial Records Department on the "ownership of the Island of Kachitivu", further corroborated India's historical claim to the island, highlighting Katchatheevu's ownership by the Ramnad king (Fraser & Tytler 1922).

The first colonial efforts to demarcate the fisheries' line between India and Ceylon dates back to 1920 (De Silva 2008). On October 24, the following year, Indian and Ceylonese delegations attempted to negotiate a "Fisheries Line" to limit the overexploitation of maritime wealth. The negotiations overlapped with the question of Katchatheevu's jurisdiction. The Ceylonese side, led by the principal collector of customs, B. Horsburgh, disputed the Indian administration's claim over the island – even though it belonged to Zamindari of the Ramnad king – by marshalling evidence of the island and the St. Anthony's Church thereupon being assets of the Jaffna Diocese. It was ultimately agreed that India and Ceylon would settle on a maritime border three miles west of Katchatheevu, which then nominally placed the island in Ceylonese territories. Although the agreement had no official seal by the Secretary of State, a makeshift maritime border was born (Jayasinghe 2003, 13–15; De Silva, 2008). In fact, the Indian delegates were cautious enough to preempt that the Fisheries Line could not officially be deemed as an international boundary to avoid prejudicing "any territorial claim which the Government of Madras or the Government of India may wish to prefer in respect of the island of Katchatheevu" (Suryanarayan 2005, 73). Subsequently, the extension of the British government's lease of Katchatheevu up to 1936 effectively excluded British Ceylon from staking any claims to the island. The status quo persisted until 1947–1948 when both India and Ceylon gained independence from British colonial rule (Rajappa 2013). In 1947–1948, a lease was granted exclusively for Katchatheevu to the Dewan of Ramanathapuram, V. Ponnuswamy Pillai, by the Indian merchant Mohammed Meerasa Maraickar. This lease signified a transfer of rights over the island, once again highlighting India's jurisdiction over Katchatheevu.

The colonial Indian administration saw the territory as encompassing the "waters between the mainland [India] and Ceylon", primarily aiming to safeguard British imperial interests in the abundant fisheries of the region. The stance was consistent with latter-day "international law requirements of long and undisturbed exercise of rights and the international recognition of these rights" (Mani 1979, 380). Hence, India's legal jurisdiction over Katchatheevu was almost uncontroversially recognized until the late 1960s. Following Ceylon's independence, on February 4, 1948, affluent Lankan Tamils, and Tamils in general, underwent a major shift in their status in the newly formed state. Sinhalese majoritarianism left Tamils feeling culturally and politically disenfranchised (Oberst 1988; Hiranandani 2005, 185). Some respite came for the Tamils in 1964, when Indian prime minister Lal Bahadur Shastri and Sri Lankan prime minister Bandaranaike reached a pact to repatriate 500,000 Lankan Tamil laborers to India, although the pact never materialized due to Lankan apprehensions about the loss of Indian laborers from Sinhalese tea estates. Sri Lanka's Tamil-Sinhalese tensions climaxed in the early 1970s with the advent of the radical Sinhalese communist group, Janatha Vimukthi Peramuna (JVP), purportedly induced by North Korean influence (Manoharan 2006, 22). Between April 5 and June 30, 1971, the group orchestrated an armed insurgency against Bandaranaike's Ceylonese United Front government. Since 1972, following the new Lankan constitution and Buddhism's official political status in the nation, Lankan policies have tended to impair existing frictions between the Tamil and Sinhalese communities, essentially contributing to the emergence of the Tamil militant organization Liberation Tigers of Tamil Eelam (LTTE) (Kearney 1978). Given their Indian roots, Tamils were suspected as political allies of India in its alleged intentions of Sri Lanka's colonization.

In 1971, when Bandaranaike solicited India's help in suppressing the Lankan agitation, the Indian navy answered by ordering its Western fleet to watch Colombo and its adjacent ports. Gradually, India's intervention became undesirable to both the Lankan establishment and the JVP, alongside the increasing involvement of US and Russian naval forces in the Indian Ocean. At international forums, such as the Non-Aligned Heads of State Conference in Cairo in October 1964, the Lusaka Conference of Non-Aligned States in September 1970 and the Singapore Conference of Commonwealth Prime Ministers in January 1971, Sri Lanka and India favored the declaration of the Indian Ocean as a "zone of peace", leading to the United Nations General Assembly ratifying the proposal in December 1972 (O'Neill & Schwartz 1974). Katchatheevu's cession to Sri Lanka, two years later, was, in no uncertain terms, a political goodwill gesture to retain Lankan reciprocity –also bearing in mind the fact that Lankans believed Katchatheevu contained petroleum reserves (Vivekanandan 2001, 78). in recent times, Sri Lanka has reportedly rented land around Katchatheevu and Needantheevu to Chinese corporations. For the Indian government, which was preoccupied with the problem of homeless Lankan Tamils, Katchatheevu possessed no strategic importance (Raghavan 2012, 146). To "mandarins in Delhi", it was virtually a "barren rock" in "midsea far from being worth fighting for with a friendly country" (Vivekanandan 2001, 78). When the issue of the "Kachcha Thivu [*sic*] Island Dispute" was first raised in the Lok Sabha, in 1956, Indian prime minister and minister for external affairs Jawaharlal Nehru immediately vetoed the Government of India or the Government of Ceylon "coming into conflict over a tiny little island", even as he reminded the parliament that, either way, Katchatheevu would remain in the ownership of the Ramnad king (*Lok Sabha Debates*, April 14, 1956, 2220–2222). For Nehru, the island was insignificant for national prestige. However, for the Ceylonese, it was an ideal territory for implementing *uti possidetis juris* – the legal principle of having colonial boundaries as boundaries of postcolonial sovereign nation-states. While India was of the view that the island should be jointly administered, Ceylon's demands grew stronger, although Katchatheevu was too insignificant to appear on most maps. In 1968, the Colombo-based newspaper *Sun* carried the contentious byline: "Ceylon Government takes over Kachcha Thivu [*sic*]'. It catalyzed renewed "controversy between India and Ceylon regarding the political status of the island", and, ultimately, obliged the Indian government "to abjure the ambivalent attitude it had taken during the last decade on the issue and to expedite a solution of this question in consultation with Ceylon" (Phadnis 1968, 783).

When the Indira Gandhi administration agreed to the cession of the island, the opinions of Indian Tamil fisherfolk were ignored. Meanwhile, Indian voices celebrated the cession to "destroy the canard that India behaves overbearingly towards its small neighbours" despite having what the Government of India believed to be "an unassailable case" that it was apparently pleased to renounce for pursuing "harmonious relations with Sri Lanka" (*Indian and Foreign Review*, 24). It was held by Indian commentators that the Sethusamudram project, which, by now, had been debated in the Indian parliament for about 20 years, could not suffer on account of the cession. As far back as 1974, rumours of a possible Chinese investment in Sri Lanka and, particularly, Katchatheevu were afloat, but the Indian euphoria dismissed it as "fantastic", hoping to focus instead on Lankan reciprocity of the goodwill act (*Indian and Foreign Review*, 24). However, the Katchatheevu crisis was far from resolved. According to Article 5 of the cession agreement, Indian fishermen and pilgrims were permitted to "enjoy access to visit Katchatheevu" without being required by Sri Lanka to obtain travel documents or visas for these purposes ("Agreement between the Government of India and ...", 1974) as also announced in the Lok Sabha by the Indian external affairs minister on July 23, 1974 (*Lok Sabha Debates*, July 23, 1974, columns 197–201). Article 6 of the agreement states that "the vessels of Sri Lanka and India will enjoy in each other's waters such rights as they have traditionally enjoyed therein" (*Lok Sabha Debates*, July 23, 1974, columns 197–201) – a principle that would be later disregarded by naval authorities of both nations. Nevertheless, a 1976 agreement by the two nations comprising an exchange of letters between the foreign ministries would complicate matters. Accordingly, the Indian fishing boats and fishers were no longer permitted to "engage in fishing in the historic waters, the territorial sea and the exclusive economic zone of Sri Lanka", and Lankan fishers and boats were to follow suit with respect to Indian "historic waters, territorial sea and the exclusive economic zone of India, without the express permission of Sri Lanka or India, as the case may be" ("Exchange of Letters", 40). Anticipating this kind of imbroglio, the Tamil political party DMK had objected to Katchatheevu's cession in the Lok Sabha, back in July 1974, terming it an "unholy and disgraceful act of statesmanship, unworthy of any government" (*Lok Sabha Debates*, July 23, 1974, column 187). According to an Indian legal precedent from the Berubari Union Case (1960) ("Berubari Union and . . . vs Unknown" 1960). India's apex court had opined Indian territories could not be handed over to other nations without amending the Indian constitution – a clause that was overlooked in the cession of Katchatheevu. In the decades to follow, while Tamil Nadu would go on to see the 1976 agreement as illicit – especially since it was transacted during the Indian Emergency when the Tamil Nadu assembly remained suspended – the Lankan view would refuse to buy the narrative that India "gifted" Katchatheevu to its littoral neighbor "through goodwill" for fostering "bilateral relations" (Jayasinghe 2003, 1).

Even so, until about the outbreak of the Lankan civil war (1983–2009), Tamil fishers could visit Katchatheevu without being castigated by the Sinhalese navy (Jayasinghe 2003, 79). Arrested Tamil fishers used to be released by the Lankans after routine investigations and minor seizures – a regimen that was reciprocated by the Indian navy in cases of arrested Lankan fishers. As the Lankan civil war became more intense, and as the LTTE began acquiring popular support from Tamil organizations and political elites, the Tamil Eelam naval wing ("Sea Tigers") became more aggressive, inducing the Lankan navy to adopt more stringent measures not only against suspected LTTE forayers, but also against straying Tamil fishers, thus turning the Palk Bay fisheries into Sethusamudram's "killing waters" (Gupta & Sharma 2007, 94).

Nucleus of Fishing Disputes

The Indo-Lankan international maritime boundary line stretches over 400 kilometers, spanning the Gulf of Mannar in the south, the Bay of Bengal in the north and the Palk Bay in the center. The maritime boundary lies about 12 nautical miles from India's Rameswaram Island and 35 nautical miles from Sri Lanka's Mannar Island. Being home to five marine turtle species, over 300 marine algal species, nearly 600 piscine species, a dozen seagrass species and several mangrove species (Salagrama 2014), the 15,000-square-kilometer littoral territory is an extensive ecological asset traversing the international boundary. Despite being Lankan territory today, Katchatheevu's importance to India cannot be underestimated as nearly 1% of the gross domestic product (GDP) and over 5% of India's agricultural GDP derive from its fisheries (Government of India 2019). The nearly one-million-strong population of Tamil fishers occupies approximately 20% of India's fisheries labor force (Central Marine Fisheries Research Institute, *Marine Fisheries Census* 2010). An ontological quality of Tamil fisher life lies in what André Wink has termed (in the context of Indian Ocean historiography) as the region's "hydrological instability" (Wink 2002, 439). In the process, Katchatheevu's aquapelagic nature is ontologically qualified by the environmental disruption of human life as a praxis.

A vivid example of the environmental instability of the region comes from the village of Ramakrishnapuram, near Rameswaram. The unstable currents around Adam's Bridge often alter the course of pelagic fishes, thus altering the destination of the fishing lines on which they are eventually caught (Salagrama 2014, 36). This underscores an essential quality of aquapelagos – that they are neither stable spatial nor geographical entities but are more profoundly fluid in both literal and metaphorical senses. Meanwhile, like ecological temperaments, the ideological doctrines of India and Sri Lanka, also run counter to the Gulf of Mannar's fragile liminality and its fishing communities. Around the Sethusamudram region, "the boundaries between legal and illegal, licit and illicit, are often blurred and the nested scales of local, national, regional, and global no longer hold tight" (Mezzadra & Neilson 2013, 236). In the psychogeographical imagination of Tamil fisherfolk, Sri Lanka still constitutes the old Tamil country, not least because of the shared Tamil ethnicity between fishers on both sides of the international border – a legacy dating back to the Chola Kingdom of circa 1100 AD that included northeastern Sri Lanka with Polonnaruwa as the local capital. In fact, nearly 40% of Indo-Lankan commerce is still operated through Tamil Nadu, while only a third of it occurs as formal trade (Manoharan & Deshpande 2018, 74). Katchatheevu, the uninhabited island, "refuses to be subjected to human intervention and confronts people with its indifference even alienness –and remains resistant in spite of human attempts to discipline and picture it in particular desired ways" (Tripathy 2008, 138). All the same, the precarious cross-border forays and arrests of Tamil fishers on both sides of the maritime boundary – with Katchatheevu as the nucleus – sketch and resketch the patchy littoral border, thereby engraving an ambivalent vestibule between cartographic boundaries and borderless homelands (Gupta 2007).

The piscine discourse around Katchatheevu cannot be ignored by its historians. This is not only on account of the fact that an increasing number of Indian fisherfolk foraging in Lankan waters have themselves been victims of the Lankan navy's aggressive marine policy that is often dubbed as "shoot first and question later" (Manoharan & Deshpande, 79). Rameswaram fisherfolk are said to be paid special incentives by owners of trawlers and motorized vessels for overnight forays in Lankan seas (Sharma 1999) in a scheme that inexorably normalizes the risk to life and property. Nearly 75% of fishers' conflicts with Lankan naval personnel revolve around fishermen from Rameswaram, followed by those from Jagadapattinam, Kodikarai and Kottaipattinam. These accidental foragers are led into Lankan waters predominantly due to outdated gear and navigational intelligence, besides the lure of shrimp that are virtually exhausted on the Indian side but still abound in Lankan waters near and around Jaffna. One major reason for the depletion of piscine resources on the Indian side is the otherwise risky business of trawling, which has now led Lankan fisherfolk to protest against the monopolization of the Sethusamudram's fisheries by Tamil bottom trawling operations (Scholtens, Bavinck & Soosai 2012, 87–95). The Indian fishing industry was boosted with the advent of trawlers during the 1960s, thanks to an Indo-Norwegian alliance aimed at promoting intensive fishing and exports from the Palk Bay and Gulf of Mannar (Srinivas 2021). Trawling, traditionally an operation suited to temperate waters, is generally harsh on tropical marine ecologies, especially those with species more complex, numerous and considerably smaller than those of colder littorals. Trawling is particularly hazardous for fish and invertebrates, leading to a chain reaction of ecological imbalances, besides the indiscriminate overexploitation of fortuitous catch comprising seahorses, seabirds, mammals, turtles and piscine juvenilia (Stiles, Stockbridge, Lande & Hirshfield 2010; Kelleher 2005).

The Palk Bay and the Gulf of Mannar, with its exceptionally entangled species and "phenomenal inter-species interaction" (Kurien 2017), were ill-suited for trawling operations, yet fishing demands overrode ecological concerns. Indian fisheries expanded in the 1980s with the arrival of advanced gear and fleet, which catapulted the industry into a "pink gold rush" (prawning) or "blue revolution" (Kurien 1978; Bavinck 2001). Subsequently, Indian fisheries became more and more privatized, given increasing incentives by the state of Tamil Nadu and motorization of artisanal vessels, which led to the normalization of Indian Tamil fishers foraging in Lankan seas (Johnson

& Bavinck 2004; Paramasivam 2006). Coinciding with the early years of the Lankan civil war, this period marked increased sorties by Indian fishers in Lankan waters in accordance with their perceived piscine resources in the Sethusamudram region (Subramanian 2007). Back in 1980, Tamil Nadu had around 2,300 trawlers with over 6,000 nets operating in the Sethusamudram region (Central Marine Fisheries Research Institute 1981). Around 1985, around 1,500 Indian trawlers, largely comprising Tamil fishers, were seen to operate around the international boundary (Government of Tamil Nadu 1986) alone, with the Palk Bay being the principal piscary until the end of the 20th century. While in the early 1980s, the Palk Bay supplied one-fourth of Tamil Nadu's total piscine produce, by 1992 the number escalated to over a third and, by the end of the 20th century, it had grown closer to 40% of the total Tamil fishing wealth (2007, Gupta & Sharma 2007, 121). Thus, by 2010, as many as 6,000 Tamil trawlers were found to be fishing in the Palk Bay region (Central Marine Fisheries Research Institute 1981; Central Marine Fisheries Research Institute 2010, 4). Meanwhile, the Lankan fisheries declined (Vijayan 2008). The piscine produce in the Jaffna Peninsula dropped from about 49,000 metric tons (1983) to a little over 2,000 metric tons (2000) (Siluvaithasan & Stokke 2006, 244). Concurrently, the produce in other districts of Tamil Nadu also deteriorated, while the Rameswaram district itself went on to control nearly 5,500 vessels in the Sethusamudram region, with nearly half of those fishing in Lankan marine territory (Vivekanandan 2001; Adams 2015). At present, Tamil Nadu contributes nearly 1,000 artisanal boats, 12,500 motorized vessels and 6,000 trawlers (nearly half of which fish in the Palk Bay alone) (Prasada 2021; Vincent 2020; Nath 2022). The state's fishing schemes seem to further normalize the principle of "non-compliance" and being more "profitable" (Shekhar & Prakhar 2020), rather than awaiting a long-delayed bilateral consensus regarding the Palk Bay fisheries – a delay that ultimately has deleteriously affected both the livelihoods of local fishers and the ecology.

The political and ecological question of the Palk Bay fisheries is also closely linked to the caste question. As widely recognized, artisanal fishers are both a technologically and an economically disempowered community, and, especially in Tamil Nadu, fishers from marginalized castes are more vulnerable to becoming collateral victims of the bilateral imbroglio (Béné 2003; Salagrama 2006; Bavinck & Johnson 2008; Weeratunge, Béné, Siriwardane et al. 2014; Béné, Arthur, Norbury et al. 2016; Olsen, Kaplan, Ainsworth et al. 2018). Official Indian records suggest that during the Lankan civil war (1983–2009), nearly 250 Indian fishermen died due to either the direct or the indirect actions of the Lankan navy. Hundreds of other Indian Tamil fishers were penalized with arrests and property losses owing to inadvertent forays across the maritime boundary. Even after the end of the Lankan civil war, Indian fishers continue to be arrested by Lankan authorities. In 2012, about 200 Indian fishers were arrested; in 2013, a little less than 700; a little less than 800 in 2015; and by July 2016, nearly 250 ("Attack on Indian Fishermen" 2015). More than 1,350 Indian fishers and 250 fishing vessels were seized by Lankan authorities between 2015 and 2018 ("Release of Fishermen in Custody of Sri Lanka" 2018). In 2020, during the global pandemic, about 80 Indian fishers were arrested. In December 2021, even at a time when the International Monetary Fund was seeking to reverse Lankan economic recession, along with Indian relief, the Lankan navy arrested about 70 Indian Tamil fishers and detained 21 trawlers from the Sethusamudram region due to alleged illicit fishing in Lankan marine territory ("As more Indian fishermen are detained by Sri Lanka" 2021). Importantly, Sri Lanka's Regulation of Foreign Fishing Boats Act, which was enacted in 1979, was revised in 2018 to deter fishing vessels from straying into Lankan waters. This allowed Lankan authorities from imposing fines of anything from 6 to 175 million SLR – further imperiling Tamil Indian fishers for whom accidental maritime border crossings have become entrenched as an extension of their precarious identities. In early 2022, the Lankan administration received a bail amount of INR 10 for arrested Indian fishers ("Sri Lankan court asks 13 Indian fishermen to pay Rs 1 crore"' 2022). Since a large number of Tamil trawling operations are controlled not by local Tamil fisherfolk but by wealthy industrialists, political leaders and media personalities, artisanal fishers remain inequal partners in the equation (Vivekanandan 2014). Furthermore, since the Palk Bay fisheries crisis is a recurring electioneering issue for both Tamil and Lankan political parties (Kapoor 2018), while neither side wins, the poor fishers of both territories lie endangered.

The Uneasy Interregnum

In 2008, towards the end of the Lankan civil war, Tamil Nadu Chief Minister J. Jayalalithaa appealed to the Supreme Court of India to review the 1974 and 1976 agreements that had distressed Tamil fisheries and claimed the lives of fisherfolk on both sides of the border. Her appeal was followed by her petition to the Indian prime minister, Narendra Modi, on the same matter ('Press Release No. 351' 2014). Recently, Tamil Nadu Chief Minister M.K. Stalin has also presented a memorandum to Prime Minister Modi, concerning Katchatheevu ("Demands made in memorandum" 2022). Meanwhile, in 2014, Mukul Rohatagi, the former attorney-general for India, opined that India would have to practically resort to war in order to retrieve Katchatheevu, since the island's cession was not contested between the governments of India and Sri Lanka but was a matter of internal federal disagreements between the state of Tamil Nadu and the Indian government (Janardhan 2014). While for Tamil politicians, Katchatheevu has been an emotional issue, the Government of India continues to maintain official restraint owing to its neighborhood-first policy (Joshi 2022) and also because the matter is sub-judice before the Supreme Court of India ("Katchatheevu Island" 2019). The entangled nature of Katchatheevu – especially given its shifting aquapelagic, political, strategic and folk dimensions – now exists in an uneasy interregnum. Despite the studied silence of the Indian government, the Indian prime minister's recent references to the island suggests that its contested historiography, which is due to its aquapelagicity, is never far from the cultural imagination of Indian (and by extension Lankan) civil society. Both India and Sri Lanka are party to the Vienna Convention on the Law of Treaties (1969), which compels them to honor the Agreements of 1974/76. The convention's article 56 explains that neither nation can unilaterally withdraw from the agreements without ratification by the other as prescribed by article 65(1) (Vienna Convention 19; 22). Furthermore, the withdrawal of either party from the agreement will require a resolution under the UN Charter's article 33, with the likelihood of intervention by a mediating state or the International Court of Justice, as accorded by article 37 ("Pacific Settlement of Disputes" 2021). It is far from adequate, however, to reduce the histories and possible routes of negotiations around Katchatheevu to merely geopolitical contingencies. On the one hand, the island begs a substantive history of its own, which is not necessarily ordained by state nationalisms of both territories nor old-school Marxist dialectical materialism that might simplify it as a saga of battles between poor fishers and powerful state actors. The more sustainable way out of this interregnum perhaps lies not in state-based solutions – for instance the shared administration of Katchatheevu, as some experts have proposed – but rather a bilateral rebuilding of civil society consensuses on the contested memories revolving around the island. As a critic of Indo-Lankan relations once humorously remarked, Lankans "may not always live happily with the Indian state, but they seem to live happily with India's national poet" (Nandy 2006, 3500) Rabindranath Tagore, who is said to have composed the musical scores for the national anthems of India and Sri Lanka. India is unlikely to make a political demand for Katchatheevu, and Sri Lanka is unlikely to offer it back. This purely geographical and political premise emphatically fails to acknowledge the aquapelagic essence of Katchatheevu. The relationship of Indian and Lankan fishers to the island far exceeds its piscine utility. It needs to be evaluated instead in terms of "aquapelagic belonging", which extends to "the sea, marine and air environment, the island and the spaces in-between bring about emotional dispositions which would appear to form different patterns of relationships compared to elsewhere" (Hayfield

& Nielson 2022, 209). As the notion of aquapelago helps us recognize, the legends of navigation, foraging and perhaps even precarity of fishers on both sides of the border are central to the scheme of their identity and belonging, which does not preclude relations with non-human others – from vessels to corals, from seaweeds to anemones, from the depletion of shrimp to the extinction of dugongs.

As is argued of aquapelagos, in general, "neither cartography nor Google Earth services (etc.) can identify an aquapelago on their own – analysis of human inhabitation of space is central" (Hayward 2012a, 2) In the case of Katchatheevu, in particular, the precarity of human habitants, in the form of fisherfolk, constitutes the ontology of the aquapelagic belonging of the larger scope of life and mobility in the Gulf of Mannar and Palk Bay region, whose historicity cannot be summarized purely in anthropocentric terms but should ideally extend to the effects of the Anthropocene on the non-human. The least that can be said is that histories like that of Katchatheevu require rigorous foregrounding by Island Studies scholars. This recognition may lead to an amplification of the island's description – in Indo-Lankan civil societies, in particular, and by island observers, in general – beyond its geographical and geopolitical denominators. Eventually, a more robust history of the island, one that conceives of it in an expanded sense as an aquapelago, cannot overlook precarity as fundamental to the aquapelagic belonging of its human and non-human denizens. This aquapelagic turn, which may also be, in fact, a turn towards affective historiography, may well be an important exit not only to an uneasy geopolitical interregnum, but also to Katchatheevu's narrowed interpretive potential – and that of the Indian Ocean in general – whose hermeneutic fluidity and multidirectional memories lie dormant due to colonial intellectual legacies (Joseph 2019, xi–xii). A case is due to be made for Katchatheevu as not merely a bilaterally contested geography but as one of the Global South's several islands entangled in frictional geo-politics. Islands such as "Bolghatty Island, Vypin Island, Fort Cochin Island and the archipelago of islands across the Kerala coast of South India [with] their histories of miscegenation and multi-religious syncretisms", are "increasingly flashpoints of contested political identities" like the Sunderban Islands of West Bengal and Bangladesh and the precarity of Rohingyan refugees, "who seek to find safety amidst the already overcrowded mudflats of Bangladesh" (Joseph 2021, 5). Historians and political observers have generally attempted to ask who Katchatheevu belongs to. The point, however, is to interpret it in various ways, while reexamining the question of ownership and the assemblages of human, non-human and commemorative relationships that constitute that ownership with due regard for the hermeneutic biases that the Anthropocene's geopolitical utilitarianism casts on our understanding of this entangled aquapelago in the guise of humanistic realism.

References

Agreement Between the Government of India and the Government of the Republic of Sri Lanka on the Maritime Boundary in the Gulf of Manaar and the Bay of Bengal, New Delhi, 23 March 1976. 1994. *India, Bilateral Treaties and Agreements: 1976–1977*. New Delhi: Policy Planning and Research Division, Ministry of External Affairs, Government of India, 30–33.

As more Indian fishermen are detained by Sri Lanka, hard questions need to be asked in the Palk Strait. 2021. *The Indian Express*, December 23.

Attack on Indian Fishermen: Q. No. 1970. 2015. Rajya Sabha, Ministry of External Affairs, Government of India, August 6.

Demands made in memorandum submitted by Tamil Nadu Chief Minister M.K. Stalin to Prime Minister Narendra Modi. 2022. *The Hindu*, June 17.

Exchange of letters constituting an Agreement Between the Government of India and the Government of Sri Lanka on the Wadge Bank Fisheries. New Delhi, 23 March 1976. 1994. *India, Bilateral Treaties and Agreements: 1976–1977*. New Delhi: Policy Planning and Research Division, Ministry of External Affairs, Government of India, 39–44.

Katchatheevu Island: Sir Creek Issue. 2019. *Lok Sabha Unstarred Question No. 1824*. New Delhi: Ministry of External Affairs, Government of India.

Pacific Settlement of Disputes Chapter VI of UN Charter. 2021. *United Nations Security Council*. https://www.un.org/securitycouncil/content/pacific-settlement-disputes-chapter-vi-un-charter.

Press Release No. 351: Text of the D.O. letter dt. 2.7.2014 addressed by Selvi J Jayalalithaa, Hon'ble Chief Minister of Tamil Nadu to Shri Narendra Modi, Hon'ble Prime Minister of India, New Delhi. Chennai: Director, Information and Public Relations, Chennai.

Release of Fishermen in Custody of Sri Lanka: Q. No. 1619. 2018. Rajya Sabha, Ministry of External Affairs, Government of India, December 28.

Sri Lankan court asks 13 Indian fishermen to pay Rs 1 crore each for bail, OPS seeks help from EAM Jaishankar. 2022. *Times Now*, April 13.

Adams, Manju. 2015. Indo-Lanka fishery issues: Traditional and human security implications. *Proceedings of 8th International Research Conference*. Colombo: General Sir John Kotelawala Defence University, 65–70.

Anand, R. P. 1968. The Kutch Award. *India Quarterly*, 24(3), 183–212.

Baldacchino, Godfrey. 2016. Diaoyu Dao, Diaoyutai or Senkaku? Creative solutions to a festering dispute in the East China Sea from an 'Island Studies' perspective. *Asia Pacific Viewpoint*, 57(1), 16–26.

Banerjee, Prathama. 2020. *Elementary aspects of the political*. Durham and London: Duke University Press.

Bavinck, Maarten. 2001. *Marine resource management: conflict and regulation in the fisheries of the Coromandel coast*. New Delhi: Sage Publications.

Bavinck, Maarten and Johnson, Derek S. 2008. Handling the legacy of the blue revolution in India-social justice and small-scale fisheries in a negative growth scenario. *American Fisheries Society Symposium*, 49(1), 585–599.

Béné, Christophe. 2003. When fishery rhymes with poverty: a first step beyond the old paradigm on poverty in small-scale fisheries. *World development*, 31(6), 949–975.

Béné, Christope, Arthur, Robert, Norbury, Hannah et al. 2016. Contribution of fisheries and aquaculture to food security and poverty reduction: assessing the current evidence. *World development*, 79, 177–196.

Berubari Union v. Unknown. 1960. AIR 1960 SC 845, 1960 3 SCR 250 (Supreme Court of India).

Central Marine Fisheries Research Institute. 1981. Marine Fisheries Information Service. *Indian Council of Agricultural Research*, 30, 19.

Central Marine Fisheries Research Institute. 2010. *Marine Fisheries Census 2010, Tamil Nadu*. New Delhi: Department of Animal Husbandry Dairying and Fisheries.

Chatterjee, Arup K. 2022. Lord Ram's Own Sethu: Adam's Bridge Envisaged as an Aquapelago. *Shima*, 16(1), 94–114.

Chatterjee, Arup K. 2024. *Adam's Bridge: Sacrality, Performance, and Heritage of an Oceanic Marvel*. London and New York: Routledge.

De Silva, Sanath. 2008. Sharing maritime boundary with India: Sri Lankan experience. Paper presented at the Working Group meeting of the Regional Network for Strategic Studies Centers (RNSSC) on WMD and Border Security Issues held from 12–15 October in Istanbul, Turkey. Washington DC, USA: US Defence University.

Dunnett, Oliver. 2021. Imperialism, technology and tropicality in Arthur C. Clarke's geopolitics of outer space. *Geopolitics*, 26(3), 770–790.

Field, A. 1903. The Legend of Adam's Bridge. *The American Antiquarian and Oriental Journal*, 25, 39–40.

Fraser, W.K. and Tytler. 1922. Report of the information reproduced from the Imperial Records Department. File number 327-G/29, Proposed Delimitation of the Gulf of Manaar and Palk Strait with a View to Safeguard the Marine Fisheries off the Madras Coast, Question of Ownership of the Island of Kachitivu, 1929, Foreign and Political Department, General Branch, Government of India, National Archives of India.

Government of India. 2019. Press Release: Fish Production and Consumption, January 8. https://pib.gov.in/newsite/PrintRelease.aspx?relid=187305

Government of Tamil Nadu. 1986. *A Census of Tamil Nadu Marine Fishermen*. Chennai: Directorate of Fisheries.

Gupta, Charu. 2007. Bonded bodies: coastal fisherfolk, everyday migrations, and national anxieties in India and Sri Lanka. *Cultural Dynamics*, 19(2–3), 237–255.

Gupta, Charu and Sharma, Mukul. 2007. *Contested coastlines: Fisherfolk, nations and borders in South Asia*. New Delhi and Abingdon: Routledge.

Hayfield, Erika A. and Pristed Nielsen, Helene. 2022. Belonging in an aquapelago: Island mobilities and emotions. *Island Studies Journal*, 17(2), 192–213.

Hayward, Philip. 2012a. Aquapelagos and aquapelagic assemblages: towards an integrated study of island societies and marine environments. *Shima: The International Journal of Research into Island Cultures*, 6(1), 1–11.

Hayward, Philip. 2012b. The constitution of assemblages and the aquapelagality of Haida Gwaii. *Shima: The International Journal of Research into Island Cultures*, 6(2), 1–14.

Hiranandani, Gulab M. 2005. *Transition to eminence: The Indian Navy 1976– 1990*. New Delhi: Ministry of Defence (Indian Navy).

Indian and Foreign Review. 1974. July 15, 11(19), 24.

Janardhan, Arun. 2014. Explained: An island marooned. *The Indian Express*, December 12.

Jayasinghe, W.T. 2003. *Kachchativu and the Maritime Boundary of Sri Lanka*. Pannipitiya: Stamford Lake Publication.

Johnson, Derek and Bavinck, Maarten. 2004. *Social justice and fisheries governance: The view from India*. Amsterdam: Centre for Maritime Research (MARE), Amsterdam School for Metropolitan and International Development Studies, UvA, Rural Development Sociology, Wageningen University.

Joseph, May. 2019. *Sea Log: Indian Ocean to New York*. London and New York: Routledge.

Joseph, May. 2021. Nomadic identities, archipelagic movements and island diasporas. *Island Studies Journal*, 16(1), 3–8.

Joshi, Urja. 2022. Assessing Indian-Sri Lankan ties in choppy waters: The fisherman issue vis-à-vis international law. *The Leaflet*, December 23. https://theleaflet.in/assessing-indian-sri-lankan-ties-in-choppy-waters-the-fisherman-issue-vis-a-vis-international-law/.

Kapoor, Ritika. 2018. Making fishing in Palk Bay "safe." *Observer Research Foundation*, June 5. https://www.orfonline.org/expert-speak/making-fishing-in-palk-bay-safe/

Kearney, Robert N. 1978. Language and the rise of Tamil separatism in Sri Lanka. *Asian Survey* 18, 521–534.

Kelleher, Kieran. 2005. *Discards in the World's Fisheries an Update*. Rome: Food and Agriculture Organization of the United Nations.

Kurien, John. 1978. Entry of big business into fishing, its impact on fish economy. *Economic and Political Weekly*, 13(36), 1557–1565.

Kurien, John. 2017. Healing the Sea. *The Indian Express*, July 14.

Lok Sabha Debates. 1956. April 14. New Delhi: Lok Sabha Secretariat.

Lok Sabha Debates. 1974. July 23. New Delhi: Lok Sabha Secretariat.

Mani, V.S. 1979. India's maritime zones and international law: A preliminary inquiry. *Journal of the Indian Law Institute*, 21(3), 336–381.

Manoharan, Nagaioh. 2006. *Counterterrorism Legislation in Sri Lanka*. Washington: East-West Center Washington, Policy Studies 28.

Manoharan, Nagaioh and Deshpande, Madhumati. 2018. Fishing in the troubled waters: Fishermen issue in India–Sri Lanka relations. *India Quarterly*, 74(1), 73–91.

Mezzadra, Sandro and Neilson, Brett. 2013. *Border as Method, or, the Multiplication of Labor*. Durham: Duke University Press.

Nandy, Ashis. 2006. Nationalism, genuine and spurious: Mourning two early postnationalist strains. *Economic and Political Weekly*, 41(32), 3500–3504.

Nath, Akshaya, 2022. Tamil Nadu govt calls for retrieval of Kachchativu island to end detention of Indian fishermen by Sri Lanka. *India Today*, April 13.

O'Neill, Robert and Schwartz, David N. 1974. The Indian Ocean as a "Zone of Peace" In T.T. Poulouse (ed.), *Indian Ocean Power Rivalry*. New Delhi: Young Asia Publications, 1974, 177–189.

Oberst, Robert C. 1988. Federalism and ethnic conflict in Sri Lanka. *Publius: The Journal of Federalism*, 18(3), 175–194.

Olsen, Erik, Kaplan, Isaac C., Ainsworth, Cameron et al. 2018. Ocean futures under ocean acidification, marine protection, and changing fishing pressures explored using a worldwide suite of ecosystem models. *Frontiers in Marine Science*, 5, 64.

Ong, Brenda M.Q. and Perono Cacciafoco, Franceso. 2022. Pedra Branca off Singapore: A historical cartographic analysis of a post-colonial territorially disputed island. *Histories*, 2(1), 47–67.

Paramasivam M. 2006. FRP boat boom in Tamil Nadu, India. *Bay of Bengal News*, September, 10–12. http://bobpigo.org/uploaded/bbn/sep_06/pages10-12.pdf

Paranavitana, S. 1958. The god of Adam's Peak. *Artibus Asiae. Supplementum*, 18, 5–78.

Phadnis, Urmila. 1968. Kachcha Thivu: Background and issues. *Economic and Political Weekly*, 3(20), 783–788.

Prasada, Lanka. 2021. Indian poaching in Lanka's waters: Going round in circles for 5 decades. *The Sunday Times*, November 14. https://www.sundaytimes.lk/211114/sunday-times-2/indian-poaching-in-lankas-waters-going-round-in-circles-for-5-decades-461777.html

Raghavan, V.R. 2012. Internal Conflicts: Strategic Overview. In V.R. Raghavan (ed.), *Internal Conflicts: Military Perspectives*. New Delhi: Vij Books, 15–176.

Rajappa, Sam. 2013. Why this double standard?' people of Tamil Nadu ask the PM. *The Weekend Leader*, 4(36), September 6.

Ramaswamy, Sumathi. 2004. *The Lost Land of Lemuria: Fabulous Geographies, Catastrophic Histories*. Berkeley and Los Angeles: University of California Press.

Salagrama, Vengkatesh. 2006. *Trends in Poverty and Livelihoods in Coastal Fishing Communities of Orissa State, India* (No. 490). Rome: Food & Agriculture Organization of the United Nations.

Salagrama, Vengkatesh. 2014. *A Livelihood-Based Analysis of Palk Bay*. CMPA Technical Report Series No. 01. New Delhi: Deutsche Gesellschaft für Internationale Zusammenarbeit (GIZ) GmbH.

Scholtens, Joeri, Bavinck, Maarten and Soosai, A.S. 2012. Fishing in Dire Straits: Trans-boundary incursions in the Palk Bay. *Economic and Political Weekly*, 87–95.

Sharma M. 1999. In risky waters. *Frontline*, 16(19) (September 24), 65–70.

Shekhar, Raj and Prakhar, Astutya. 2020. The Indo-Lankan fishing water conflict vis-à-vis United Nations Convention of the Law of the Seas. Centre for *Maritime Law* (National Law University Odisha, Cuttack) June 24. https://cmlnluo.law.blog/2020/06/24/the-indo-lankan-fishing-water-conflict-vis-a-vis-united-nations-convention-of-the-law-of-the-seas/

Sherif, Shereen. 2013. Negotiating postcolonial spaces: A study of Indo-Sri Lankan fishing disputes. *International Studies*, 50(1–2), 145–164.

Siluvaithasan, Augustine S. and Stokke, Kristian. 2006. Fisheries under fire: Impacts of war and challenges of reconstruction and development in Jaffna fisheries, Sri Lanka. *Norsk Geograsfisk Tidsskrift—Norwegian Journal of Geography, 60*(3), 240–248.

Srinivas, Sitara. 2021. The Palk Bay Dispute: Trawling, livelihoods and opportunities for resolution. *Social Political and Research Foundation, 11*(9), 1–8.

Stephen, Johnny. 2015. *Fishing for Space: Socio-Spatial Relations of Indian Trawl Fishers in the Palk Bay, South Asia, in the Context of Trans-Boundary Fishing*. Doctoral dissertation, Universiteit van Amsterdam.

Stiles, Margot L., Stockbridge, Julie; Lande, Michelle and Hirshfield, Michael F. 2010. Impacts of bottom trawling on fisheries, tourism, and the marine environment. *Oceana*, 1–11.

Subramanian, Ajantha. 2007. Community, place and citizenship. In M. Rangarajan (ed.), *Environmental Issues in India: A Reader*. New Delhi: Dorling Kindersley, 444–453.

Suryanarayan, V. 2005. *Conflict Over Fisheries in the Palk Bay Region*. New Delhi: Lancer Publishers.

Tripathy, Jyotirmaya. 2008. Territory and identity: The Indian experience. *South Asian Review, 29*(2), 133–157.

United Nations. 2023. BBNJ Agreement on Marine Biodiversity of Areas beyond National Jurisdiction. https://www.un.org/bbnjagreement/en.

Van Aert, Peter. 2016. The Beagle Conflict. *Island Studies Journal, 11*(1), 307–314.

Vienna Convention on the Law of Treaties, 1969. (1980). United Nations.

Vijayan, A.J. 2008. An overview of the marine fisheries and fishers in and around Rameswaram, Tamil Nadu (Unpublished Draft Report). In C. Gupta and M. Sharma (eds.), *Bonded Bodies: Coastal Fisherfolk and National Anxieties in South Asia*. New Delhi: Routledge, 1–9.

Vincent, S. 2020. Palk Bay fishing problem requires Indo-Sri Lankan joint-governance. *Maritime Affairs: Journal of the National Maritime Foundation of India*, 16(2), 71–88.

Vivekanandan, V. 2001. *Crossing Maritime Borders: The Problem and Solution in the Indo-Sri Lankan Context*. Chennai: International Collective in Support of Fishworkers, International Ocean Institute, 76–89.

Vivekanandan, V. 2014. *From the margins to centre stage: Consequences of Tsunami 2004 for the fisher folk of Tamil Nadu*. Presentation to the Planning Commission of India, New Delhi, October 18.

Weeratunge, Nireka, Béné, Christopee, Siriwardane, Rapri et al. 2014. Small-scale fisheries through the wellbeing lens. *Fish and Fisheries*, 15(2), 255–279.

Wink, Andre. 2002. From the Mediterranean to the Indian Ocean: Medieval history in geographic perspective. *Comparative Studies in Society and History*, 44(3), 416–445.

9

LENAPEHOKING/NEW YORK

An Estuarine Aquapelago

Philip Hayward and May Joseph

Historical Visions

New York, known to the Indigenous Lenape as Lenapehoking, is Turtle Island's[1] largest metropolitan archipelago. Lenapehoking's settler colonial history of island becoming and archipelagic forgetting is gradually rediscovering the area's lost practices of aquapelagic living. The transformation of the city's understanding of itself as an archipelagic environment has accelerated under the pressures of climate adaptation. Lenapehoking was colonized by the Dutch, then the British, and today it is a settler colonial city gradually grappling with its Lenape pasts. Settler colonialism (Wolfe 2006) began erasing Indigenous ways of interfacing with the archipelago's shorelines in the 1600s and today the estuarine city is exploring new infrastructural dynamism across its many shores. Climate change has dramatically transformed New York's coastline from a "wasted waterfront" (Silber 1996) to an increasingly amphibious landscape. This chapter traces some key ideas that spurred the coastal thinking of New York at a time when the city's island interconnectivities were still nascent.

In 2009 the City of New York celebrated the 400th anniversary of Captain Henry Hudson's exploration of the mouth of the river that now bears his name; an event that led to the initial Dutch settlement of New Amsterdam and thereby the origins of contemporary New York. Against a backdrop of Native American communities calling attention to the conflicted history of Dutch colonization (Vettikal 2022) and the Hudson River Valley's settler colonial narratives (Boztas 2024), the Museum of the City of New York curated an exhibition entitled "Mannahatta/Manhattan". The exhibition was organized by the Brooklyn-based Wildlife Conservation Society

DOI: 10.4324/9781003569534-9

(WCS) and was accompanied by a website and a large, impressively illustrated volume entitled *Mannahatta* (after the original Indigenous name for the island) (Sanderson 2009). Deploying an inventive combination of historical sources, contemporary interpretation and visualization techniques, the exhibition documented the manner in which the island's original inhabitants, the Lenape, existed within the landscape (Sanderson & Browne 2007). In sublime contrast to the island's present-day existence as a vertiginous agglomeration of skyscrapers, a congested grid road system and expanding park spaces in the new era of climate adaptation, with a population of over 8 million, Sanderson estimated that the island was populated on a seasonal basis by between 300 and 1,200 individuals in the early 1600s. Sanderson (2009, 10) also contended that:

> If Mannahatta existed today as it did then, it would be a national park – it would be the crowning glory of American national parks. [it] had more ecological communities per acre than Yellowstone, more native plant species per acre than Yosemite, and more birds than the Great Smoky Mountains National Park. Mannahatta housed wolves, black bears, mountain lions, beavers, mink, and river otters; whales, porpoises, seals, and the occasional sea turtle visited its harbor. Millions of birds of more than a hundred and fifty different species flew over the island annually on transcontinental migratory pathways; millions of fish – shad, herring, trout, sturgeon and eel – swam past the island up the Hudson River and in its streams during annual rites of spring. … Oysters, clams and mussels in the billions filtered the local water; the river and sea exchanged their tonics in tidal runs and freshets fueled by a generous climate; and the entire scheme was powered by the moon and the sun, in ecosystems that reused and retained, water, soil, and energy, in cycles established over millions of years.

Sanderson's characterization of early-17th-century Manhattan as a palimpsestic biosphere emphasizes the power sources and mechanisms underlying Mannahatta's ecological assemblage. Deploying georeferencing, spatial data sets and landscape ecology, he provides a vivid projection of the significance of Manhattan's estuarine location for its (former) ecological diversity that merits quotation in full:

> History, geography, and climate all set Mannahatta up to be a biological success, but what makes Mannahatta wealthy beyond imagination is its crowning position atop an estuary… By definition, estuaries are the places where the land and sea come together, and the result is like currency, both productive and variable. Freshwater rivers, like the Hudson and the numerous streams that are her sources and tributaries, discharge

nutrients to fertilize the water, and cut the saltwater with fresh flow. As the seasons turn, the amount of freshwater swells and diminishes, and as the days and nights pass, the tide rises and falls. The competing traffic of freshwater and seawater and the washing of water over land creates a small sea in the glacially evacuated harbor, with layers of warm ocean water lying on top of the cold, fresh stuff. Sea-grass beds take root where the water is shallow enough for light to reach the bottom, beaches and dunes form along the windward shore, and salt marshes thrive in pro-tected corners. The estuary is the motor, the connector, the driver, the great winding way, the central place that gathers all the old neighbor-hoods together and makes the rest possible.

(Sanderson 2009, 143)

Sanderson's muir webs use historic maps of Mannahatta to visualize key ideas germane to the study of the estuarine aquapelago (Sanderson 2007, 550). The reassembled scanned images foreground the interaction of layers and types of water in the estuarine environment as crucial to its biodiversity. He underscores the manner in which the encircling estuarine waters aggre-gated and regulated the various ecological neighborhoods on and around the island. The estuarine location is identified as engendering the island's biodi-versity and, thus, the conditions that proved conducive to Lenape habitation.

Sanderson's rich computational geography of Mannahatta's estuarine ecology foregrounds the interdependent complexity of the island's pre-colo-nial ecologies. In a series of images in the Museum of the City of New York exhibition and within Sanderson's book that contrast GIS renderings of his-torical landscapes with images of similar areas of the contemporary island dominated and topographically altered by built structures, an approach to Mannahatta's ecology opens up a rethinking of its archipelago structure. The reassembled digitized description of layers of the historical estuarine environment, including terrestrial and marine biota, are seminal to how the city now imagines its ecological and marine habitats. Central to this scaf-folded ecological understanding of the landscape prior to Henry Hudson's arrival is the concept of the aquapelago and the assemblages that produce it.

The Aquapelago

As elaborated in Chapter 1 of this volume, aquapelagos are assemblages of terrestrial and aquatic elements that are utilized and navigated in man-ners that are fundamental to a resident social group's livelihood and their senses of identity and belonging. But not all aquapelagos are equivalent. We can identify a spectrum of assemblages. At one end are ones that are low level in terms of their human population density and the intensity of their human community's utilization of natural-environmental resources.

Typically, such uses do not cause severe disruptions to the landscape, marine flows, environmental chemistry (and thus, terrestrial and marine habitats and the variety and interactive dynamics of the organisms that inhabit the area). The nature of Lenape inhabitation of Mannahatta in the pre-contact period, when a population of around 300–1,200 inhabited an area with 51.5 kilometers of coastline and 87.5 square kilometers of terrestrial surface on a seasonal basis, exemplifies this type of aquapelagic assemblage. Such assemblages can be characterized as essentially sustainable. At the other end of the spectrum are assemblages where the intensity of resource use and disruption to the environment mark them as both short-lived and unsustainable. These aspects result from two (often overlapping) factors:

a)   an intensification of exploitation of particular types of marine species for commercial gain in regional, national or international markets (and/or significant increases in the populations of localities that cause similar surges in demand),
b)   a growth of human population that severely disrupts the locale's landscape, seascape and environmental chemistry (and thus, terrestrial and marine habitats and the variety and interactive dynamics of the organisms that inhabit the area).

In contrast to sustainable types of assemblage, unsustainable assemblages have a predictable narrative of resource decline and environmental disruption that principally varies in terms of key components and/or the duration of their cycles. In the case of contemporary Manhattan, the exponential population growth and intensification of resource extraction were interrelated and the particular activity of oyster harvesting and its support industries provides a crucial historical hinge between the early Indigenous inhabitation of Mannahatta and the later, massively disruptive, industrialization of the city's foreshore.

In his book *The Big Oyster* (2007), Mark Kurlansky provided a vivid overview of the marine/terrestrial interface that was crucial to the pre- and post-contact populations of Manhattan. In allusive contradistinction to the city's 20th-century nickname of "The Big Apple", the book provides a detailed historical account of the centrality of the Hudson River's oyster beds to the development of the island. Kurlansky's study facilitates an understanding of the island's modern history in terms of two distinct types of aquapelagic assemblage, the artisanal and the industrial. The artisanal is a low-impact and sustainable system. On Manhattan this was generated by a population that numbered around 50,000 in the early 1800s that had access to a large area of oyster habitat that was not significantly disturbed by human construction activities. Its transition to an industrial system resulted from the rapid increase in the island's population, which rose to

500,000 by 1850 and peaked at 2.3 million in 1910 (Angel & Lamson-Hall 2014). As Kurlansky relates, the industry attempted to service the escalating population by intensifying its local operations. At its peak, in 1880, the waters around Manhattan were dominated by an industrial aquapelagic assemblage that employed a wide range of personnel in gathering, delivering, shucking, wholesaling and retailing the mollusk (alongside others employed in constructing and maintaining oyster barges, docking facilities, etc.). This created multiple nodes of interaction between estuarine and terrestrial environments. The industrial operation delivered a staggering 700 million oysters per year to local markets at its height.

As extensive and intensive as the industrial assemblage was, it was also completely unsustainable. The period of peak production around 1880 was followed by a complete collapse of the oyster industry that coincided with the island's highest population density, creating something of a "perfect storm" for the local oyster industry. A cluster of factors combined to eventually obliterate the mollusk around the New York–New Jersey harbor estuary. Chief amongst these were massive stock depletion, dredging operations and land reclamation, which decreased the area of oyster habitats. Unregulated pollution further impacted the quality and toxicity of the remaining New York harbor oysters. These factors rapidly terminated the local industry as less compromised stock was sourced from farther afield, with the final New York harbor oyster bed closing in 1927 (Billion Oyster Project).

In tandem with the above factors, the city's rising prominence as a port, serving both its own internal populace and the broader region, also severely impacted the oyster's coastal habitats. As Raymond Gastil (2002) has detailed, from the mid-19th to the mid-20th centuries, shipping facilities came to dominate the southern half of Manhattan and the East Side, requiring coastal flats to be constructed to allow direct shipping access. Foreshores and wetlands were also drained and/or built over in order to allow ships' cargoes to be transferred to land and dispatched via rail and road. With its oyster industry in rapid decline, the city lost a major element of its livelihood connection to its adjacent waters. In just over 300 years the complex and bountiful ecological diversity that Sanderson emphasized as the island's "crowning" aspect in the early 1600s had all but vanished.

The use of large tracts of Manhattan's coast for port facilities created an industrial buffer zone around the island's interior that severely inhibited public access to shorelines and thereby to coastal waters. This buffering restricted the construction of the type of coastal frontage housing that could maintain a visual-imaginative identification of the city as an island surrounded by waterways (see Swaminathan, 2014 for a discussion of this aspect with regard to the development of Mumbai). Faced with such exclusions from its waterfront; parks – and the city's iconic Central Park, in particular – became a focus for the population's leisure activities. In many ways,

Central Park can be imagined as a "walled garden", hemmed in by the high-rise buildings that surround its four sides on an island whose residential areas were constricted by the commercial dock facilities that surrounded its shores. The decline of Manhattan's shipping industry from the 1960s on, accelerated by the introduction of container shipping suitable for large, automated dock facilities, brought little relief and even ossified the situation, with waterfront facilities variously being locked off and/or reassigned for new coastal roadways that also blocked public access to the foreshore and to its (severely depleted) fishery (Gastil 2002).

Sanderson and Kurlansky's research indicates that between the mid-1800s and early 1900s, Manhattan's aquapelagic orientation declined into insignificance. The parallel collapse of the oyster industry, the rise and fall of Manhattan as a shipping hub, and the continuing ascendancy of the island as a global commercial center reflected a retreat from its islandness and the island's coastal fringe. New York's forgetting of its aquapelagic histories were brought into clear focus by a major weather event that impacted the lower Hudson estuary in 2012, Hurricane Sandy.

Rising Currents: Hurricane Sandy and Reengagement with the Aquapelagic

Hurricane Sandy was a Category Three storm that crossed the North American coast close to Atlantic City (200 km south of New York City). The hurricane produced a storm surge up the Hudson that hit Manhattan, adjacent islands and coastal locations on 29 October 2012, causing major flooding of low-lying areas and major disruption to transport routes dependent on tunnels and subterranean facilities. The same narrowing estuarine location that Sanderson (2009, 142) characterized as "the connector [...] that gathers all the old neighborhoods together" provided a funnel for storm surge inundation and damaging wave action. Storm surges around Manhattan rose as high as 3.85 meters above normal tide levels, causing extensive flooding. While fatalities were limited (with 72 deaths in the United States being directly linked to the hurricane), damage was extensive, with 650,000 homes being either destroyed or damaged. The financial cost of the disruption has been estimated at just under $50 billion overall and $19 billion in New York City alone (Blake et al. 2013). If New Yorkers had largely forgotten their estuarine location and the fundamental interconnectivity of their land areas with the marine environment that surrounded them, Sandy's impact catastrophically reminded them of it. The hurricane's impact underlined the primacy of the geo-physical space New Yorkers inhabit, its "chiasmatic idiosyncrasy" (Maxwell 2012, 23) and the broader "onto-story" of their hydrological location, i.e. of "nature doing what it does" (Bennett 2010, 116–119).

The design and materiality of the city's built foreshore was a major factor in the damage that resulted from the storm surge. As Grizzle and Coen (2013, 327) have emphasized:

> The urbanized shorelines of New York Harbor have long been hardened by seawalls, bulkheads, docks and other structures ... but in large measure these are structures that are essential for a working port. The hardened shorelines were designed to withstand the hydrodynamics and other conditions typical of busy harbors ... [not] to withstand extreme storm events.

In *Fluid New York* (Joseph 2013), May Joseph points out that Hurricane Sandy redefined New York's aquapelagic identity. During the Dutch and British periods, New York's islandness was pronounced by an entirely water-based form of transportation. There were no bridges or tunnels. Boats defined how people knew their environs. The archipelago was both water-bound and aquapelagic, accentuated by its prominence as the world's largest maritime port by the middle of the 19th century (Seitz & Miller 1996). The archipelago's ring-like formation of little islands strung like pearls along the New York harbor and the East River informed the aquapelago's interrelational history of designating different islands for different social functions during the colonial and subsequent era of urban growth (Joseph 2019).

The aquapelagic interrelationality of the New York archipelago sharpened its peripheralization of New York's minor islands following independence. As the Hudson River gained ascendancy as the primary engine of the archipelago through the 19th century, the delicate small island infrastructures bolstering the soaring modernity of Manhattan expanded (Seitz & Miller 1996)). There were islands for incarceration (Rikers Island); islands for the mentally ill and orphans (Roosevelt Island, Ward Island) (Cudahy 1997); islands for the quarantined (Castle Clinton); islands for the burial of the indigent (Randalls Island and Hart Island); islands for immigrant communities replicating the seafaring lifestyles of their European heritage that would become landfilled over time (such as City Island, Edgemeer Island and Broad Channel); military islands (such as Staten Island and Governors Island); alongside islands for navigational machinery (Liberty Island); islands for the port industry (the Red Hook archipelago); islands for immigration (Ellis Island); islands for segregated leisure (Coney Island); as well as islands designated for public housing of the poor (the Rockaways) during Robert Moses's era. These living island cultures and ecologies coexisted in relation to each other. For Manhattan to invent its daring reach towards the skies, these peripheral islands had to exist as sites of the underbelly of modernity (Seitz & Miller 1996). Along these peripheral small islands ringing Manhattan were housed the sick, the incarcerated, the dead, the abandoned, the destitute, the immigrant and the infectious, including the licentious and excessive of Coney Island (Joseph 2019). These aquapelagic interdependencies coexisted in relation to the great city of New York, which was Manhattan until 1898 (Williams 2014), when Brooklyn became officially incorporated into the City of New York. The islands of Staten Island and Queens were incorporated later.

In 1904, 147 ferry boats operated around New York City, servicing all five boroughs and including New Jersey (Glowinski 2019). By 1925, there were only a dozen routes servicing New York and New Jersey. By 1945 the only functioning ferry service was the Staten Island ferry (Cudahy, 1990). This dramatic decline of an earlier aquapelagic history, followed by the heavy hand of New York's bridge and tunnel master Robert Moses, accelerated a historic forgetting of the city's archipelagic geology. Robert Moses laid the groundwork for a car-based, automotive logic that cut the coast off from the islanders (Berman 1988). Moses's monumental infrastructures severed the "ecological DNA" of New York's aquapelagic knowing from its ways of living and commuting. The ferries connecting New York's archipelago shut down (Cudahy 1990). The ports fell into disrepair as ground travel gained prominence. The waterfronts were abandoned derelict spaces and cars defined the new mid-20th century of Robert Moses's New York and the best views of the city were staged for the automobile (Caro 1975). New York archipelago's estuarial thinking eroded, leading to a 20th-century land-based terrestrial logic that would create major infrastructural problems for the island city in the era of 21st-century rising seas. People forgot where the original shoreline of 1609 was, and, in 2023, New Yorkers were baffled as to why the city was getting flooded catastrophically in locations away from the shoreline (Villafane 2023). The answer lies in the city's ecological forgetting of its water sources and the voracious land filling of its archipelago structure.

Anticipating this crisis, architects, engineers and urban designers of New York began raising concerns about the potential impact of storm surges and a general increase in sea levels (largely if not conclusively attributed to human-produced global warming) from the 1990s on. This prompted a number of artists, architects, theorists and planners to explore potential futures for New York that addressed its estuarine location in a more integrated and interactive manner. One particularly influential study was the "On the Water/Palisade Bay" project (henceforth "OTW/PB"), led by Guy Nordenson, Catherine Seavitt and Adam Yarinsky from Princeton University's School of Architecture, in 2007. Taking as its starting points New York and New Jersey's location in the estuary of the Hudson River and the phenomenon of rising sea levels, the study aimed to both lower the risk of flooding events and create a new, more integrated interface between metropolitan built environments and estuarine spaces. The team identified their proposals as being based on a "soft", flexible and adaptive infrastructure:

which aims to synthesize solutions for storm defense and environmental enrichment along the coast. It is an adaptable solution that adjusts to varying climatic conditions and urban demands by balancing environmental, technical, and economic priorities. Our goal is to layer these priorities throughout the harbor zones to not only create a comprehensive storm defense system but also to provide new places for recreation, agriculture, ecologies, and urban development. By arraying these activities on the water, the bay becomes a regional centre, and the city refocuses on the body of water it surrounds.

(Nordenson, Seavitt & Yarinsky 2009, 12)

In contrast to "hard", monumental infrastructural developments (such as those constructed for ports and for modern ship berthing, in particular), the team's approach engaged with New York's estuarine harbor space as "a fluid body with a porous boundary" where the "figure-ground relationship of the water and the land constantly changes as it is subject to forces ranging from diurnal tides, floods and dry seasons, and modes and intensity of use" (Nordenson, Seavitt & Yarinsky 2009, 20). In this manner, the team focused on the city's immediate coastal fringe and, significantly, the "intensity of use" of that area. The latter aspect identifies the livelihood activity core to the constitution of aquapelagic assemblages in environments that are constantly subject to change. These require ingenious and adaptive patterns of human use of spaces that are "soft" in the sense of being both "fluid" and "porous". The final project report proposed three initiatives:

- the construction of "an archipelago of islands and reefs along the shallow shoals of the New York–New Jersey Upper Bay to dampen powerful storm currents as well as to encourage the development of new estuarial habitats";
- the revitalisation of the waterfront "by designing a broad, porous, 'fingered' coastline which combines tidal marshes, parks, and piers for recreation and community development";
- the enactment of "zoning formulae that adapt efficiently in response to the impact of storms in order to increase community resilience to future natural disasters" (Nordenson, Seavitt & Yarinsky 2009, 22).

Substantially inspired by the example of the OTW/PB study, Barry Bergdoll, curator of architecture and design at New York's Museum of Modern Art (MoMA) in 2007–2013, secured funding for an initiative entitled "Rising Currents". This comprised the research, visualization and proposal of five projects for New York's waterfront that aimed to combat problems of rising sea levels and integrated metropolitan and estuarine spaces in a productive and sustainable manner. The projects essentially involved the reimagination of modern-day New York in more aquapelagic terms. The selected projects were developed by different teams in consultation with Bergdoll and others and were presented for public scrutiny and feedback at an exhibition at MoMA in October 2010,[2] together with an accompanying book (Roberts 2011). Significantly, and appropriately for the venue, the project teams were given the brief of producing work that could inspire and facilitate a reenvisioning of New York (re-)integrated with the Hudson estuary.

The Rising Currents projects addressed the particular environmental issues of specific areas and proposed the remediation of former "dirty" industrial sites into clean, sustainable facilities and local projects intended to minimize the impact of inundation. Two of the projects, for Zones 2 and 4, are of particular relevance to this chapter's discussions by dint of proposing "acupunctural" planning work in an estuarine environment. The notion of "acupunctural urbanism" is closely associated with the work of Jaime Lerner (2014), former mayor of the Brazilian city of Curitiba. The premise of acupunctural urbanism is that the city is effectively a living organism with particular pressure points that can be addressed by highly targeted interventions that can alleviate stress and/or malaise within the broader system. Drawing on this, Marco Casagrande (2012) has promoted the concept of local "biourban acupuncture", which promotes local green initiatives in urban contexts. The latter include so-called guerrilla gardens and the preservation (and enhancement) of natural grow-back. These are far from incidental initiatives; they serve to change people's senses of urban space and of their potential to reconfigure that space, creating new environments from the detritus of metropolitan history. Manhattan has a shining example of the latter. The vegetation that established itself between the 1980s and the 1990s on Manhattan's disused elevated High Line railroad, on the island's West Side, inspired the creation of the acclaimed High Line Park, which opened in 2009. Diller, Scofidio and Renfro's visionary use of adaptive reuse techniques alongside Piet Oudolph's biodiverse landscape design had a major impact on the use, community perception and sense of space in the Meatpacking District. The resounding success of adaptive reuse along the High Line led to a wider "greening" of the surrounding area of the city (David & Hammond 2011) although not without significant impact on real estate costs and related gentrification (Jacobs 2017).

The Zone 3 and 4 projects explored acupunctural approaches in ways that tacitly recognize that the humans who are implicated into aquapelagic spaces necessarily interact with a wide range of actants. The Zone 4 project had historical resonance in that it explicitly acknowledged the decline of New York's oyster industry under the pressures of excessive harvesting and habitat destruction and involved a concept that the landscape architect Kate Orff (2011) has termed "oyster-tecture". The project was designed to remediate Brooklyn's Gowanus Canal and its area of entry into the Hudson around the Bay Ridge Flats. The canal is widely regarded as one of the most polluted waterways in the United States as a result of seepage of residual pollutants from earlier industrial gas plants in the area. Orff's vision was to create a network of oyster beds and underwater rope scaffolding for shellfish that would create an offshore storm surge barrier, a new habitat for marine life, a filtering system for the polluted canal emissions and a livelihood and recreational space for locals (Orff 2011, 90–99). Orff's "soft" infrastructural proposal involved the growing of new structures within the harbor by stimulating shellfish clusters in the former location of (once commercially exploited) oyster beds. Her project reflected a more general concern to regenerate oyster beds around New York that developed in the 1990s and early 2000s based on an awareness of the need to restore the city's marine and terrestrial habitats. A decade after Orff's initial concept of Living Breakwaters, the New York harbor will see the infrastructural additions of Orff's ideas in collaboration with Murray Fisher and Pete Malinowski's Billion Oysters Project (BOP). In 2024, Orff's design interventions to remediate the water and marine environs of the New York harbor will be realized by the installation of a living oyster-tecture with oyster shells from the BOP. These living habitats of oyster islands are intended to regenerate and "re-aquapelagise" the estuarial health of the Hudson-Raritan ecology (Klinenberg 2021).

Similar strategies have been advocated by environmental artist Mara Haseltine, who has undertaken a number of projects over the past fifteen years to facilitate and publicize the restoration of oyster populations around New York (Haseltine, 2011). These include designing a solar-powered artificial reef structure, which was installed in McNeil Park in Queens in 2007; designing optimal shapes for the clustering of oyster populations in artificial oyster bed structures; and teaching a course at Manhattan's New School in 2008–2012 entitled "The Art of Urban Oyster Restoration" (see Levitt 2009). Haseltine's work has been premised on the concept of "Geotherapy", a practice that encourages humans to counter Anthropocene impacts in part by stimulating and deploying natural processes. Orff's structural skeins utilize "soft" infrastructures in a manner that evokes comparison to Haseltine's work and that extends her vision into metropolitan planning. The symbolic and applied aspects of returning to a specific species (i.e. oysters) to rebuild coastal marine environments and provide a protective, filtering fringe for the city today provides an ambitious landscape design undertaking unfolding across New York's archipelago. A new aquapelagic New York is beginning to emerge across its complex hydrology, connecting the borders of islands to each, making the estuary's extensive water bodies aware of each other's interdependencies (Kadinsky 2016).

Envisaging the "Aqueous City"

The most ambitious project in the Rising Currents initiative was Zone 3 "New Aqueous City", envisaged by Eric Bunge and Mimi Hoang for the Verrazano Narrows between Staten Island and Brooklyn. This project went further than others in the "Rising Currents" initiative by imagining a *network* of new structures of various kinds in the estuary. Bunge and Hoang (2011, 100) state:

> This project blurs the boundaries between land and sea, extending the city into the water. Habitable wave-attenuating piers (supporting housing, public leisure areas, and protected wetlands) provide docking points for a network of biogas ferries, and an archipelago of man-made islands connected by inflatable storm barriers encourages silt accumulation, fostering natural resilience against storm surges. At the same time, the water is extended into the city, which is punctuated by a network of infiltration basins, swales, and culverts that absorb storm runoff and function as parks in dry weather.

The Zone 3 project was conceived as both a new aqueous settlement in its own right and a venture that could significantly modify the character of the greater New York area by creating a new aqueous/aquapelagic fringe. In contrast to Manhattan's boxed-in interior spaces, the proposed aqueous settlement was envisaged as requiring habitation in a manner that is more fundamentally connected to and interactive with the region's estuarine environment. In addition to a series of striking images depicting a radical reenvisioning of a metropolitan interface with an estuarine environment, the project included a clearly articulated team statement in which Bunge and Hoang (2011, 100) describe their vision of the aqueous city growing quasi-organically from:

- archipelagos interconnected with dynamically inflatable barriers;
- piers extending land-based transportation out into the water; and
- bridgelike structures that would allow the suspension of lightweight housing over the water.

The architects have described how they envisaged these "infrastructural armatures" as "seeds" that could not only accommodate ecological growth, but also lead to new technologies and "evolving mind-sets" (Bunge & Hoang 2011, 108). The "soft" patterns of aggregation and interaction envisaged here mirror the organic structures in coastal waters that have allowed humans to inhabit and profitably exploit aquapelagic assemblages (Manhattan's 19th-century oyster industry being a case in point). The radical reimagination of the city's foreshores and estuarine environments that Bunge and Hoang propose would require its developers and inhabitants to negotiate and interact with the estuarine locale in a flexible and interactive manner, implicating themselves in the region's and planet's ongoing "onto-story". Key to this vision is a sense of the creation of a particular form of domicile, referring to a person's place of residence or ordinary habitation to which they have some form of continuing identification and/or commitment. The particular domicile that Bunge and Hoang propose involves humans actively inhabiting the coastal fringe and utilizing its waters (rather than simply exploiting them from terrestrial bases).

The imagination of a new aqueous city extends out into the waters of the Hudson but stops short of a reimagination of the complexity of those waters and how their former ecological diversity might be regenerated or refigured. As Sanderson (2009, 143) has elaborated, prior to metropolitan development, the Hudson was a highly complex, layered, three-dimensional space that altered in seasonal cycles, with varying proportions of salt and fresh water. The potential for "acupunctural urbanism", let alone "biourban acupuncture", in a modern-day estuarine environment that is bounded by hard infrastructure and polluted by various runoffs is staggeringly complex. Aquapelagic spaces may be constituted by human agency (Hayward 2012a, 2012b), but they are not *determined* by that agency. The vitality of non-human elements comes into play, more specifically their capacity "not only to impede or block the will and designs of humans but also to act as quasi agents or forces with trajectories, propensities, or tendencies of their own" (Bennett 2010, iii). Projects such as Bunge and Hoang's new aqueous city represent important steps towards envisaging and developing new types of coastal metropolitan settlements. More broadly, these speculative infrastructures propose a harmonious relationship between dense urban aggregations and adjacent waterways, even if they are not so much environmentally restorative as innovative.

Aquapelagic Futures

Visions of an aquapelagic city are starting to redefine coastal New York. Climate change has forced the island city to embrace a variety of infrastructural plans that interface with its terra-aqueous ecology. Interpretive architectures such as the amphibious Little Island, the new rewilded piers around Manhattan and along the East River's Brooklyn and Queens shores as well as the changing marsh and dune landscapes around Roosevelt Island, Governors Island, Rockaways Islands and Red Hook, are part of this new aqueous vernacular. The remediation measures along superfund sites such as New Town Creek and Gowanus Canal have dynamized the coast around these toxic sites, leading to revived placemaking activities such as kayaking, canoeing, music shows, nature walks and social gathering (Zimring 2024). The East Side Coastal Resiliency (ESCR) stretching across the Lower Manhattan coast is redesigning a new landscape with pedestrian bridges, coastal land banks and absorbent infrastructures. The ESCR seeks to build resiliency while also providing access to the waterfront. Bioswales, rewilding, afforestation and absorbent architecture are part of current New York archipelagic techniques to live with water. For the architectural firm BIG, which is developing the New York coastline through the concept of the "Big U", the concept of the "adaptive archipelago" as "a series of strategies for adapting various existing and artificial archipelagos to our planet's changing climate" is critical to their design philosophy (Erdman 2023).

David Erdman, director of the Center for Climate Adaptation at Pratt Institute which is part of the New York Climate Exchange on Governor's Island, emphasizes that archipelagic thinking is key to New York's survival. "What is our strategic position in the built and natural environment?" Erdman asks. Building adaptation in Red Hook and Governors Island has been central to Erdman's pedagogy. Recent local initiatives such as the New York Climate Exchange's exhibition curated by Erdman on Governors Island titled *IA: Island(ing) Adaptations*, foregrounds this urgency. The thrust of the exhibition is that "infrastructural capacity and policy needs to be discussed and understood for archipelagos". The *IA: Island(ing) Adaptations* brochure further states:

> Over one tenth of the world's population currently lives on islands. By 2030 more than 50% of the planet's fastest growing urban areas are on and/or contain islands within their border. Archipelagos, formed by clusters of islands, are among the most densely populated and climatologically vulnerable areas of the planet and include a wide array of cultures and races. They harbor both historic and prospective future practices of climate adaptation.
>
> *(Erdman 2023)*

This chapter has reviewed and synthesized various aspects of the reenvisioning of Manhattan and the wider New York area apparent in the OTW/PB and Rising Currents projects and subsequent initiatives. The projects propelled urgent aquapelagic possibilities through new integrated relationships with the city's estuarine environments. The implicit and explicit reference point for many of these projects is the more innately integrated *habitus* of Manhattan's pre-metropolitan populace, for whom the surrounding waterways were a crucial livelihood source.

Our discussions have sought to emphasize that these visions are timely and important, given the significant variations in sea levels that are projected to occur in this and subsequent centuries. The projects' visions of a new metropolitan aquapelago accreting around the shores of Manhattan and adjacent areas of the Hudson estuary are inspirational, for all the questions the proposals leave unanswered. The proposals' main achievement, in this context, is to have imagined how a metropolis located in an estuarine environment might engage with its environment in a more advantageously aquapelagic manner. They offer an important step towards envisaging and developing new types of coastal metropolitan settlements and more harmonious relationships between dense urban aggregations and adjacent waterways.

Projects such as OTW/PB and Rising Currents, alongside on-the-ground island-habitat–based discussions such as those unfolding at the New York Climate Exchange, suggest the potential to develop more ambitious domiciles along the aquatic fringes of metropolitan centers in a manner that might have profound implications on how human societies respond to the challenges of rising seas. Designers working on such projects are increasingly moving beyond a terrestrial model in which the marine aspect is regarded as an external factor. But while New York is beginning to acknowledge the value of such a holistic approach, it is not yet being implemented on the massive scale necessary to address the city's aquapelagic issues in the face of climate change.

Notes

1 Turtle Island is an alternative name for North America used by various First Nations communities and Indigenous activists.
2 See Bergdoll (2011) for discussion of the overall project rationale.

References

Angel, S., & Lamson-Hall, P. 2014. *The rise and fall of Manhattan's densities, 1800–2010*. New York University Marron Institute of Urban Management, Working Paper 18. https://marroninstitute.nyu.edu/uploads/content/Manhattan_Densities_High_Res,_1_January_ 2015.pdf
Bennett, Jane. 2010. *Vibrant matter: A political ecology of things*. Durham: Duke University Press.
Bergdoll, Barry. 2011. Rising -currents: Incubator for design and debate. In Roberts, Rebecca (Ed.) *Rising currents: Projects for New York's waterfront*. New York: Museum of Modern Art, 12–31.
Berman, Marshall. 1988. *All that is solid melts into air: The experience of Modernity*. New York: Penguin Books.
Blake, Erik. S., Kimberlain, Todd B., Berg, Robert J., Cangialosi, John B. & Beven, John L. II. 2013. *Tropical cyclone report – Hurricane Sandy*. National Hurricane Centre. https://www.nhc.noaa.gov/data/tcr/AL182012_Sandy.pdf
Boztas, Senay. 2024. Manahahtáanung or Manhattan? Tribal representatives call for apology for Dutch settlement of New York. *The Guardian,* May 15. https://www.inkl.com/news/manahahtaanung-or-manhattan-tribal-representatives-call-for-apology-for-dutch-settlement-of-new-york-city
Bunge, Eric., & Hoang, Mimi. 2011. New Aqueous City. In R. Roberts (Ed.) *Rising Currents: Projects for New York's Waterfront*. New York: Museum of Modern Art, 100–109.

Caro, Robert. 1975. *The power broker: Robert Moses and the fall of New York.* New York: Vintage.

Casagrande, Marco. 2012. *Biourban Acupuncture. Treasure Hill of Taipei to Artena.* International Society of Biourbanis

Cudahy, Brian J. 1997. *Around Manhattan Island and other tales of Maritime NY.* New York: Fordham University Press.

Cudahy, Brian J. 1990. *Over and back: The history of ferryboats in New York Harbor.* New York: Fordham University Press.

David, Joshua & Hammond, Robert. 2011. *High Line: The inside story of New York City's Park in the Sky.* New York: Farrar, Strauss and Giroux.

Erdman, David. 2023. *IA: Island(ing) Adaptations.* New York: The Trust For Governors Island, Pratt Institute.

Gastil, Raymond W. 2002. *Beyond the edge: New York's new waterfront.* Princeton: Princeton Architectural Press.

Glowinski, Patricia. 2019. *The rise, fall, and rise again of the New York City Municipal Ferry System.* New York City Department of Records and Information Services. https://www.archives.nyc/blog/2019/7/29/ferries

Grizzle, Raymond E. & Coen, Loren D. 2013. Slow down and reach out (and we'll be there: a response to "Shellfish as Living Infrastructure" by Kate Orff. *Ecological Restoration, 31*(3), 325–329.

Haseltine, Mara G. 2012. Sustainable reef design to optimize habitat restoration. In Goreau, Thomas J. & Trench, Robert K. (Eds.) *Innovative methods of marine ecosystem restoration.* Boca Raton: Taylor & Francis, 245–261.

Hayward, Philip. 2012a. Aquapelagos and aquapelagic assemblages. *Shima* 6(1), 1–10.

Hayward, Philip. 2012b. The constitution of assemblages and the aquapelagality of Haida Gwaii. *Shima* 6(2), 1–14.

Jacobs, Karrie. 2017. The High Line Network tackles gentrification. *Architect.* https://www.architectmagazine.com/design/the-high-line-network-tackles-gentrification

Joseph, May. 2013. *Fluid New York: Cosmopolitan urbanism and the Green imagination.* Durham: Duke University Press.

Joseph, May. 2019. *Sealog: Indian Ocean to New York.* London: Routledge.

Kadinsky, Sergey. 2016. *Hidden waters of New York City: A history and guide to 101 forgotten lakes, ponds, creeks, and streams in the five boroughs.* New York: W.W. Norton.

Klinenberg, Eric. 2021. The seas are rising. Could oysters help? *The New Yorker.* https://www.newyorker.com/magazine/2021/08/09/the-seas-are-rising-could-oysters-protect-us

Kurlansky, Mark. 2007. *The Big Oyster: A molluscular history of New York.* New York: Vintage.

Lerner, Jaime. 2014. *Urban Acupuncture* (3rd ed.). Washington: Island Press.

Levitt, Julia. 2009. Sculptor Mara G. Haseltine on coral reefs, biomimicry and eco art. https://www.theguardian.com/environment/2009/may/15/network-coral

Maxwell, Ian. 2012. Seas as places: Towards a maritime chorography. *Shima* 6(1), 27–29.

Miller, Stuart & Seitz, Sharon. 1996. *The Other Islands of New York City : A Historical Companion.* Norton & Company.

Nordenson, Guy, Seavitt, Catherine & Yarinsky, Adam. 2009. *On The Water/ Palisade Bay.* http://www.palisadebay.org/Chapters/01%20Introduction_lr.pdf

Orff, K. 2011. *Oyster-tecture.* In Barry Bergdoll (Ed.) Rising Currents: Projects for New York's Waterfront (pp. 90–99). Museum of Modern Art.

Roberts, R. (Ed.) 2011. *Rising currents: Projects for New York's waterfront*. New York: Museum of Modern Art.

Sanderson, Eric W. & Brown, Marianne. 2007. Mannahatta: An ecological first look at the Manhattan landscape prior to Henry Hudson. *Northeastern Naturalist* 14(4), 545–570.

Sanderson, Eric W. 2009. *Mannahatta: A natural history of New York City*. New York: Abrams.

Silber, Kenneth. 1996, April. The wasted waterfront. *City Journal*. https://www.city-journal.org/article/the-wasted-waterfront

Swaminathan, Ramanathan. 2014. The epistemology of a sea view: Mindscapes of space, power and value in Mumbai. *Island Studies Journal* 9(2), 277–292.

Vettikal, Ann. 2022. The Lenape of Manahatta: A struggle for acknowledgement. *The Eye*. https://www.columbiaspectator.com/the-eye/2022/09/06/the-lenape-of-manahatta-a-struggle-for-acknowledgement/

Villafane, Matthew. 2023, October 18. Severe flooding in and around NYC: Why does it keep happening? *CBS News*. https://www.cbsnews.com/newyork/news/severe-flooding-in-and-around-nyc-why-does-it-keep-happening/

Williams, Keith. 2014, July 24. Brooklyn's evolution from small town to big city to borough. *Curbed New York*. https://ny.curbed.com/2014/7/24/10069912/brooklyns-evolution-from-small-town-to-big-city-to-borough

Wolfe, Patrick. 2006. Settler colonialism and the elimination of the native. *Journal of Genocide Research* 8(4), 387–409.

Zimring, C (forthcoming 2025). Arts and recreation as environmental activism: Reimagining Brooklyn's Newtown Creek and Gowanus Canal in the 21st Century In Philip Hayward (Ed.) *Blue-Green Rehabilitation Urban Planning, Leisure and Tourism in River Cities*. CABI.

10

WE, THE SUBMERGED

(Non)Humans, Race and Aquapelagic Relation: Notes from New York

Ayasha Guerin

The Epoch of Racial Capitalism

The aquapelago framework, with its focus on relations between terrestrial and aquatic bodies and elements, has supported scholars to confront the violent rearrangements of life and death in the Anthropocene. Proposed by Crutzen and Stoermer (2000), the Anthropocene is a concept that has been adopted by scholars, artists and media writers to refer to our geological epoch in which human agency has become a planet-reshaping force of nature. In "Geology of Mankind" (2002) Crutzen suggests that the temporal starting point of the Anthropocene coincides with the Industrial Revolution and James Watt's 1784 refinement of the steam engine. Lewis and Maslin (2015) challenged this theory, suggesting that the epoch's start is earlier: 1610, the lowest point in a decades-long decrease in atmospheric carbon dioxide measured in arctic ice cores. They argued this drop was caused by the death of over 50 million Indigenous Americans after one century of European contact, the result of exposure to diseases carried by Europeans, plus "war, enslavement and famine" (Lewis & Maslin 2015, 177). As Luciano (2015) has well summarized, the Anthropocene debates are over what kind of story can and should be told about human impact on the planet; they are attempts not only to periodize, but also to sculpt the narrative for how socioecological relations have gotten to this dire point.

I'm not convinced that naming this climate emergency after humans will provoke us towards new relations in the future. "anthropo-" comes from the Greek root for "man". It is paired with "-cene", meaning "new" or "recent", and thus introduces the quality of time. There are reasons to be wary of the Anthropocene, which is, from the start, troubled by a flattened reference to a

DOI: 10.4324/9781003569534-10

mono-humankind. Naming the Anthropocene broadly, without identifying those specific actors who are driving global pollution spills, species extinction and ballooning atmospheric gasses, is to ignore the fact that a minority of planetary actors are responsible for the devastating effects on the majority of life on earth.

As Gomez-Barris (2017, 4) warns, the Anthropocene presents a "universalizing idiom and viewpoint that hides the political geographies embedded within the conversion of complex life". Indeed, the Anthropocene is a political concept about inequity, where specificity is key. There is no universal history of human disturbance on earth – deforestation, extraction and pollution are actions done by specific actors. Furthermore, Moore (2016, 80) writes, "the anthropocene makes for an easy story because it does not challenge the naturalized inequalities, alienation, and violence inscribed in modernity's strategic relations of power and production". We might better study planetary ecological restructuring through the lens of capitalism. Moore's "Capitalocene" framework acknowledges, like Lewis and Maslin, that slavery and colonialism have had a deep impact on world ecology. Capitalism has also restructured how we are allowed to spend our time and how time is defined, not only via the relations we have with humans, but also by the relations we have with land and oceans, plants and animal life.

I have moved away from using the term "Anthropocene" altogether, rather adopting the "racial capitalocene", a term coined by Verges (2017), who adds an explicitly racial dimension to Moore's critique of the Anthropocene. The racial capitalocene framework urges us to integrate the long memory of colonialism's impact on environments with the goal of analyzing "how categories of difference, gender, class and race, have affected conceptions of nature and of being human" (Verges 2017, 77). Verges also warns that the term "Anthropocene" maintains the nature/society division dear to Western thought, which masks the fact that relations between humans are themselves produced by nature (Verges 2017, 76). Scholars who have taken up the aquapelago framework have sought to attend to this last oversight.

As an intervention, the aquapelago, introduced by Hayward (2012), offered a corrective to the strong terrestrial focus within the field of Island Studies and described regions in which aquatic spaces play a vital constitutive role in social and cultural life. Hayward's (2015) exploration of the New York aquapelago brought focus to the socioecological constructions on New York's waterfronts and between New York's islands, encouraging a reading of the relation between humans and their aquatic environments as aquapelagic assemblages where human and non-human life change and develop each other. The aquapelago is engaged with assemblage theory, popular with new materialists like Alaimo and Bennett, who attend closely to the material connections between the human and the more than human, "reconfiguring understandings of ecologies as bodily relations between the animate and inanimate" (Alaimo 2010, 2). As Tsing (2015, 22–23) writes, "assemblages are open-ended gatherings" that "allow us to ask about communal effects without assuming them" and which "show us potential histories in the making".

Bennett's (2010) *Vibrant Matter* framework, which looks to the agency of non-human matter and engages the field of relational ontology with ecology, encourages a new materialist study of "the political ecology of things". She describes all matter as "lively" and explains that while vitality is not equivalent to life, vitality exceeds human orderings and must be considered a political force (2010, viii.) Likewise, Alaimo's (2010) "trans-corporeality" framework traces the material interchanges across human bodies, animal bodies and the wider material world. Like these approaches, an aquapelagic perspective allows us to consider that eco-modernity is not just a human story, but also a story of how our actions impact our environments *to act on us*. Still, it remains important to contextualize: these actions are shaped by social signifiers. Lack of specificity about who is responsible for climate catastrophe, and the attending socioeconomic crises, obfuscates the responsibility of those with a hand in colonial capitalism's violence.

Indigenous scholars and feminist thinkers (including Dhillon 2017; Goeman 2017; Todd 2017; Tallbear 2019; Yazzie & Baldy 2018; Barker 2019) build on ancestral knowledge and anticolonial perspectives to bring together "multiple strands of materiality, kinship, corporality, affect and land/body connection," in the political ecology project of "radical relationality" (Yazzie & Baldy 2018, 2). Indigenous scholars of radical relationality do not call themselves "new materialists" as these ways of thinking are not new for their cultures. The trans-corporeal reach of this orientation, or the "inter-reflexivity" of radical relationality, directs "a struggle of decolonization premised on the accountabilities we form in lively relation to each other" (Yazzie & Baldy 2018, 2).

The question Bennett asks in the introduction of *Vibrant Matter*, "How could political responses to public problems change were we to take seriously the vitality of non-human bodies?" (2010, vii) is layered with new meaning when we consider the historical exclusion of Indigenous and Black bodies from the category of the human and the erasure of slavery's violence from colonized shorelines. As Moten (2003, 1) writes, the "history of blackness is testament to the fact that objects can and do resist". There is danger in new materialist and post-humanist perspectives that jump too quickly beyond discussions of the human because to do so runs the risk of perpetuating the marginalization and abandonment of Black, Indigenous and other racialized peoples who never graduated into the human, as it were. If, "to be excluded from the category of the human is to be subject to unthinking, banal and routine brutalizations of everyday life under capitalism (Emejulu

2022, 39–40), then discourse on and beyond the human, too, must break the banal routine brutality of racial ignorance.

Wynter's (2001) writings identify how the white colonial subject only encounters herself or himself of being human in relation to the deficiency of the Black and Indigenous subject. Building upon Wynter's critiques of the category of the human in Western subjectivity, where the invention of *Man* as the political subject of the state is made through his possessive relation to other life as property, I join a lineage of Black feminist thinkers who question desire for inclusion in this project. Wynter encourages that we craft other genres of the human, or other codes for what it means to be human, within the history of species. As Holland (2023, 30) articulates, we are trying to "get Blackness past a humanity that so signals a bankrupt status".

Work in Black Studies offers post-humanist invitations to, rather than overlook race, recognize racial signifiers as constructing the fundamental exclusionary structure of humanism. Holland clarifies (2023, 20), "what is still left at issue is not that these hierarchies exist but what manner of relation/ relating might come forward to contest them and how?" How might we divest from the human? Or, "how might we organize our community-building to remake the world outside and against the human?" And, where human is an excluding category, "what would it look like to generate relations of care and solidarity with those others outside the category of the human?" These last questions are offered by Emejulu (2022, 9) in their work *Fugitive Feminism*.

Fugitive Feminism introduces fugitivity as an audacious and dangerous act of self-liberation, where movement is a recurrent concern (2022, 57.) Adopting the fugitive feminist struggle means looking forward to "the end to humanity itself". Emejulu's theory invites us to become a fugitive from gender and from the human to reject "that genocidal binary that imperils all of us" (2022, 69). While fugitive feminism does not name the Anthropocene, it is concerned with time: a fugitive feminist is "in pursuit of the end of human time – the linear colonial time of extraction" (2022, 71). Emejulu suggests turning to practices of counter-storytelling and remembrance, liberatory care practices, where care is a dangerous act of defiance. They write, "we become fugitives from the human through the development of an ethos of care," where those "others" are recognized as fellow fugitives (Emejulu 2022, 93–94).

Care, they argue, cannot be separated from distress in the politics of the human. In fact, "care and harm are fused together, since one informs and gives meaning to the other in the same way that human and non-human co-constitute each other" (Emiejulu 2022, 27). In the racial capitalocene, care is not required for non-human others, yet appeals to humanity often beg for participation in a social relation of care. When we defy liberal humanism and exit humanity, where do we go for care? We might enter into relationships of mutual vulnerability with other non-human life. We might turn towards longstanding Indigenous ontologies that acknowledge our radical relationality with other entities like land and water, as well as with other species. For example, when Tallbear writes about "caretaking", she is thinking beyond the human; she is thinking about "caretaking as kin or as Peoples in alliance with reciprocal responsibilities to one another and to our other-than-human relatives with whose land, water, and animal bodies we are co-constituted" (2019, 36).

Sunaura Taylor's work *Beasts of Burden* (2017) writes animal studies through the lens of Disability Studies and ableism, reminding that the category of the animal has been fundamental to histories of categorization and constructions of difference, inferiority, savagery, dependency and dehumanization. She explains, "the animal, and consequently, the human, are complicated categories, socially determined rather than solely biological (ibid:19). While Emejulu locates the Black subject in the realm of the non-human, she does not explicitly engage with animals. As Holland (2023, 2) explains, "black critical inquiry up until now has sensed the uncomfortability of its alignment with the animal, and so it tended to move away from emphasizing animal life, and toward some kind of meaningful negotiation with the human as its constant interlocutor". Here, Taylor's (2017, 110) questioning holds: "How do those of us who have been negatively compared to non-human animals assert our value as human beings without either implying human superiority or denying our own animality?"

Holland's *An Other* (2023) offers another mode of engagement, arguing "the emphasis on human being can be shattered, was shattered, by African-descended practices and imaginaries and the boundary of hum/animal" (ibid: 5). If we agree that all mammalian life is animal life and that species is "an inhabitation that marks difference among, rather than between" (2017, 12), there is radical potential for caretaking (in the obligatory kinship way that Tallbear names). As Taylor notes (2017, 81), "the struggle for animals is inseparable from the struggle for the environment more broadly", a fundamental truth which connects the wellbeing of individual bodies to our total ecologies in the unfolding of climate disaster.

Emejulu's question, "what would it look like to generate relations of care and solidarity with those others outside the category of the human?" is a question that courses through my own work, and I do turn to animals in my attempt to answer. Embracing ways of thinking "the idea of Black freedom through the multiplicity of species" (Holland 2023, 30) – allows us to advance an Afropessimist critique, which is a project produced upon the negation of Black freedom but does not offer a theory of change. In my own slip from the human, I've looked to interspecies relations between Black and Indigenous peoples and the animals they've worked and lived with, to probe the mutual implications of race and species categories for governing the free

FIGURE 10.1

FIGURE 10.2

movement of the non-human. I study the conditions and possibilities for our mutual survival. I look for stories that reveal the ways we have extended attention to one another's life, echoing Emejulu's question, "How do we practice care in ways that fundamentally transform our social relations?" (2022, 9).

To Be Submerged in the Aquapelago

I share Hayward's aquapelagic lens on the New York waterfront because it is a place where colonial value systems imported racial signifiers to control and consume non-human life. I understand the transatlantic slave trade and the colonization of the Americas as an aquapelagic, interspecies history, full of important assemblages of plants, animals and bacteria. As an artist, I am thinking about how to articulate the importance of these aquapelagic assemblages, these interspecies vulnerabilities and relational histories that have been submerged by historical processes of exclusion and cultural ignorance. How might I share what I have come to know, beyond the shorelines I walk or the closed doors of academic conferences, and how might my artworks open channels of feeling into the implications of these histories and pre-colonial possibilities? In one creative work, an audio-visual workbook (or video essay,) titled *Submerged* (2001–ongoing), my voiceover reminds:

> To be submerged is to be unseen from the perspective of land. Of course, below the water, one is seen by other sea life, and other relationships are forged.

My work considers what are, and what have been, and what could be the relations between we, the submerged. Multispecies beings that have been cast, through logics of colonial capitalism, as in/unhuman.

Submergence, in this work, becomes a shared condition of a fugitive state. An overwhelming experience shared by the non-human. I want to consider the perspective we might gain if we were "to unmoor from humanity" (Emejulu 2022, 98). I layer imagery from historical archives (photographs, maps, maritime drawings, ship ledgers and other legal documents) with film footage of coastal life, anchored by a voiceover that shares interspecies narratives and theoretical work in Black and Indigenous studies. Two years after I made my first cut, I continue to return to this work and edit, adding new footage and making new connections for future screenings. In her analysis of *Submerged*, Bezan (2023, 1) describes my practice of workbooking as an *anti-hydrostasis* approach that unsettles historical oceanic/coastal archives. She explains, "as a deliberately unfinished practice, Guerin's audio-visual workbooking serves as a valuable tool for cultivating an attentive and meaningful connection to the lived resonances of coastal histories". I am grateful for this reading. My coastal methodology looks to the shoreline as spaces of military conquest and borders that define the human and exclude others from human care.

As places where the land meets the ship, shorelines have historically been sites of migration and slavery, abolitionism and fugitivity, where the ship was a portal to freedom or at least the dream of it. A focus on Black and Indigenous life in the New York marine trades pulls perspective to the ways in which work at sea was often mediated by desires for freedom on land.

In *Submerged*, I share that one reason so many Black New Yorkers lived at sea in the 18th and 19th centuries was because of the security of Seamen's Protection Certificates. Beginning in 1796, the U.S. federal government issued certificates to free, working seamen, which defined them as citizens. While on land they would not be recognized as US citizens until 72 years later, with the passing of the Fourteenth Amendment, Seamen's Protection Certificates positioned, in writing at least, a Black man's status as something equal to his white counterpart. These papers were often faked or lent to men to escape slavery in southern ports through shipping industries. Thinking with the aquapelago, I ask: what can we make of the fact that Black people had to leave land in order to be recognized as equals?

Even when thousands of Black people in New York gained their freedom in the early 19th century, they sought waged work on water in the oystering and whaling industries, which were some of the few employment options available to them. New England whaling industries reconfigured Indigenous communities' sacred relations with whales, while the unsustainable logic of extraction that exhausted whale populations was also extended to whalers who accrued large debts in times of unsuccessful hunts. On Long Island, New York, when Indigenous peoples were unable to come up with payments, they had to forfeit tribal land to the court. This introduction of debt was often the means of Native land dispossession. As Strong (2018, 125) traces, "it is not too much of an exaggeration to say that the use of Indian indebtedness to obtain land was successful, and perhaps more widespread, than the use of alcohol or fraud". This debt displaced and pushed Indigenous and Black people into new arrangements at sea.

In my study, "Black" is a racial category that includes Indigenous descendants, as many Indigenous and free Black people intermarried over three centuries of work together in the New England whaling industry. Bringing together an interdisciplinary and interspecies reading of displacement – of whales from their natural habitats and of Black and Indigenous people from their lands into the whaling industry – *Submerged* offers an interspecies analysis of the errant condition of both whales and the indebted whaler, connecting the disorder of colonial capitalism to the endangerment of all non-human life.

Spillers (1987) reminds us that enslaved subjects were taken into account as quantities of flesh – not as bodies of individuals – before they were valued in the New World. Black whalers returned to the sea under the pressure of these events. But while it was flesh which served the primary narrative

and islands everywhere - require ways of nonhumanist thinking

FIGURE 10.3

FIGURE 10.4

of their arrival, it was flesh of whales that they returned to sea to find, to recover their own narratives of survival. Such fleshy intimacies are punctuated by the fact that whaling ships and slave ships shared the same oceanic routes, and that some whaling ships became slave ships when the Atlantic whales were gone (Reilly 1993, 180).

Whales were not passive actors in this brutal history either. When attacked, they thrashed and fought back, a deadly prospect for tired, harpoon-wielding whalers in small row boats. But whales are also very intelligent and move when heavily attacked in one area. Communicating with one another through a process of echolocation, active sonar which bounces off objects, whales responded to their endangerment by using echolocation's codes to help them sense and register large moving targets to navigate new routes. I've been inspired by the metaphoric implications of echolocation. Echolocation is a sensory mode of listening, a way of "getting a sense" of the relational dynamic between the vitality of things and of locating the self via the other. As whales use echolocation, I suggest we might listen to the resonances of relation as a method. When we listen to the reverberations of historic whaling diasporas, we might ask, what connections resonate between our time and theirs? How might temporal echoes help us to better trace how the past informs the future within relational theories of oppression? Might attention extended to the survival of the non-human other bring emergent strategies for future survival?

Importantly, Taylor (2017, 207) encourages us towards "an ethic of care" that "asks how we can learn to listen to animals, and how we can help and care for them without the paternalism and infantilization that allows them to be seen as voiceless". While much of my listening has been mediated by historical records of Black and Indigenous life at sea and by naturalists' descriptions of whale behavior back then, I am also listening to the whales of now, studying their movements and ways of communication today.

In her book *Undrowned*, Gumbs (2020, 7) argues for the value of "listening to marine mammals, specifically, as a form of life that has much to teach us about the vulnerability, collaboration, and adaptation we need in order to be with change at this time". It is an approach that resonates with the invitations of Holland (2023, 7), who questions, "what if the scene of living or that thing that looks like 'endurance' were not so upsetting and the rock were to be lifted with care, in the presence of great *thoughtfulness* under a watchful eye, as kin"? Like Todd (2017, 106) who, in leaning into a sense of reciprocal responsibilities to place (rather than human beings or time,), invites an understanding of "fish-as-political citizens", my treatment of the New England whaling aquapelago attends to the failures as well as the future opportunities for "reciprocal relationality" between aquatic species.

Sharing Gumb's and Holland's orientation to mammalian kinship, in *Submerged* I move beyond focus on mutual endangerment to possibilities of mutual care. The boundary drawn between the human and animal matters, especially for those whose identification defines the border. Identifying with whales, rather than to study them as bygone resources of flesh and power, is the difference between understanding oneself as being treated *like* an animal versus being treated *as* one (Holland 2023, 21). This reading of relation deconstructs how codes of race have structured our differential experience even as we decenter the category of the human.

I bring this same lens to Black and Indigenous relationships with oysters in the New York aquapelago. In the early 17th century, the Lenapehoking estuary was filled with millions of oysters. The Lenape-cultivated reefs supplied people with a large part of their diet, especially during winter when land life hibernated (Pritchard 2007, 39) (Sanderson 2009, 106–107). European settlers soon adopted the oyster as a mainstay of their culinary culture too. The reefs would feed their growing population and also those colonies connected to New York by trade routes. Under this colonial, mercantile capitalism, in just two centuries of settlement and slavery, one of the most oyster-rich habitats in the world was exhausted.

As the port city grew and struggles for enslaved people's liberation leveraged knowledge of and access to sailing networks and the trades of sea animals, many free Black people in New York chose to work as common oystermen. In 1810, more than half (16 of 27) of the oystermen listed in the city directory were free African Americans (Hewitt 1993, 240). Unlike working a job on land, where competition with poor white laborers could be dangerous, Black oystermen could work for themselves most of the year. This not only influenced Black coastal settlement but also the destinies of the aquatic species with which their lives were entangled at sea. When New York oystermen saw their livelihoods threatened during the first oyster collapse, as natural shellfish beds began to thin in the 1810s, Black oystermen transplanted to New York small "seed" or tiny, larval "set" oysters from fertile Virginia beds (Burrows & Wallace 1999).

On one of the islands in New York's aquapelago is the community of Sandy Ground. By the south shore of Staten Island, it was one of the oldest oystering communities established by free Black people in North America. To be an oysterman required a mastery of sailing and the muscle to operate wind-powered boats. Some workers held sea skills from their lives in West Africa or the Caribbean, and Indigenous knowledge for net-harvesting mollusks was passed down through generations of enslaved people brought to the Americas (Warsh 2010). Sandy Grounders utilized basket-making skills, practicing indigenous West African cultural traditions on new land. They cut white-oak saplings, split them into strips to soak in water and wove them into bushel baskets that were used to transport oysters between reefs and the piers of Manhattan (Mitchell 1956). Over 40 African-American Sandy Grounders were listed as oystermen in the city directory by 1900 (Askins 1991, 8).

Harvests were regulated through a series of traditional customs which promoted the long-term sustainability of the beds, seen as a resource held in common for all oystermen. Common oystermen were hired by captains for part of the season, but they otherwise worked to cultivate the public, natural growth beds, where it served in their best interest to act as ecosystem stewards. And underwater property was an unsettled concept: the ownership of a planted bed was considered forfeited if the planter did not return to harvest annually. Jacques's (2017) research details how the shift from subsistence gathering to seeding and growing to the industrial extraction era of oystering coincided with the privatization of underwater commons. New legislation after 1900 imposed costly licensing, requiring the registration and taxation of oyster beds. It also refused oystermen the right to work planted beds after the end of the season. The merchants behind these changes called themselves "oyster capitalists"; they cut out the middlemen – captains – and invested in thousands of acres of harbor waters to take from. Their corporations used dredge machinery to gather oysters on powered boats, scraping at reef habitats.

Unlike those caretakers who had understood the fragile tipping points of harbor ecosystems and acted as stewards in an artisan-like culture of oystering, oyster capitalists viewed beds as short-term resources for their own profit. The final component of the attack on the customary culture of oystering was the mobilization of Jim Crow racism, which legally separated and excluded African American oystermen from opportunities. They found themselves unemployed as they faced discrimination during harvests. These companies would not buy seed or mature stock from them, as captains of the trade once had, and Black oystermen were increasingly allowed to work only as deck and shore hands (Askins 1991).

In this way, a Black and Indigenous subsistence relationship with non-human life – the New York oysters – was colonized. Meanwhile, as whales were endangered and petroleum discovered, runoff from John D. Rockefeller's Standard Oil released three million gallons of crude oil per week into the New York estuary, poisoning shellfish further (Hurley 1994, 345–346). The city's growing population overwhelmed the harbor with sewage, causing severe oxygen drops in regional waters and fish kills. All of these factors accelerated the oyster industry's collapse at the turn of the 20th century.

Despite Staten Island's important role in progressive Black history, it is now New York's most conservative and whitest borough. It is the place where Eric Garner couldn't breathe. It is the place where, during Hurricane Sandy, Glenda Moore's toddlers died because a property owner, who believed her to be a Black man knocking at his door, locked them out in the surge (Hume 2012). In the aftermath of the mother's pleas and the toddlers' drownings, the property owner told reporters he was forced to spend the night with his back against the door Moore banged on, "to prevent entry and thereby his own violation" (Sharpe 2016, 78). In an era when care is not required and extended to non-human others, it is a scene that illustrates the fallacy of a shared humanity and gendered condition.

are interconnected with histories of Indigenous and Black displacement from land.

FIGURE 10.5

Changing relations between whalers and whales reflect the importance of layering post-colonial and ecological perspectives together.

FIGURE 10.6

Confluence in Conclusion

The aquapelago framework within Island Studies is most truth-bearing when it attends to how historical, colonial relationships have shaped socioecological assemblages of hum/animal species, animate and inanimate matter's vitality, differently across racialized geographies. To advance discussions of "the political ecology of things" in Island and Ocean Studies, the fields must acknowledge the ways in which historical constructions of racial difference and racialized space have influenced these assemblages of human and non-human relationships to act upon each other. We will have to do more than name our frameworks of study to really, truly attend to the climate emergency brought on by colonial capitalism, but in a world where the Los Angeles Police Department classifies Black subjects with words "no humans involved" (see: Winter 1994) and Israeli military officials name Palestinians "human animals" (Gallant 2023, 00:12) as a license to kill them, the "racial capitalocene" is a more honest naming for the relations of this epoch to which the aquapelago framework must attend.

In my research, the aquapelago is a primary site of slavery's apparatus and a space where the afterlives of slavery emerge as places of imprisonment. For example, consider the bodies drifting in "The Boat," the Vernon C. Bain Maritime Facility in Bronx waters, an 800-bed jail barge currently used to hold inmates for the New York City Department of Corrections. While I have argued that attending to the violent rearrangements of the racial capitalocene is important, I am equally convinced that there is much to remember in turning to pre-colonial relations of the aquapelago. The fugitive possibility of escaping the violence of our time lies therein.

As we place our hopes in a post-Anthropocene world, we must look to and learn from Indigenous life, including Indigenous African life, before transatlantic slavery. Indigenous ontologies extend relation and kinship accountabilities not only to other species, but also to water, "reactivating water as an agent of decolonization as well as the very terrain of struggle over which the meaning and configuration of power is determined" (Yazzie & Baldy 2018, 1). Yazzie and Baldy write that the many Indigenous stories that foreground relationships to water show us that water is theory. They argue that "this multidirectional, multispatial, multitemporal, and multispecies theory of relationships and connections forms the terrain of decolonized knowledge production" (2018, 1). Barker's writing on water as an analytic of Indigenous feminisms likewise states that "water teaches us to be mindful of our relations with one another, including other-than-human beings and the lands and waters on/in which we live together" (2019, 6). Recognizing our shared vulnerability in water, the aquapelago framework has the potential to attend to "the intricacies and intimacies of imperial violence" (2019, 6) affecting multispecies lifeworlds.

There is guidance in remembering an epoch marked by an abundance of life and care, before the invention of Western liberal humanism and the reception of mass death. Indigenous efforts to preserve traditional aquatic cultures, wherever socioecological aquapelago research is focused, should be included in studies. Adopters of the aquapelago framework can look beyond the boundaries of our lifetime to acknowledge the genius of Indigenous peoples who have held and continue to hold reparative aquapelagic relations with the more than human. They are reminders that another world is possible.

The extractive impulse to drill without regard for ecological relation, whether in the head of a whale or the foot of a mountain, has propelled modern life into a crisis of multispecies loss and racial violence. The layers of pollution and death in New York's harbor reflect an interspecies, racial-capitalist history. The ongoing challenges facing vulnerable coastal communities in New York and islands everywhere require ways of non-humanist thinking that are responsive and responsible to the historical conditions which continue to produce precarious assemblages of being for racialized people and our kin. With the aquapelago framework, we must continue to ask, and to trace, how extractive relationships at sea are interconnected with histories of Indigenous and Black displacement from land. To plot our own fugitive routes towards survival, we, the submerged, embrace a logic of care to conceptualize being and belonging, beyond the category of the human.

References

Alaimo, Stacy. 2010. *Bodily natures: Science, environment, and the material self.* Bloomington: Indiana University Press.

Askins, W. 1991. Oysters and equality: Nineteenth Century cultural resistance in Sandy Ground, Staten Island, New York. *Anthropology of Work Review 12*(2), 7–13.

Barker, Joanne 2019. Confluence: Water as an analytic of Indigenous Feminisms. *American Indian Culture and Research Journal 43*(3), 1–40.

Bennett, Jane. 2010. *Vibrant matter: A political ecology of things*, Durham: Duke University Press.

Bezan, Sarah. 2023. Coastal methodologies: Audio-Visual Workbooking in Ayasha Guerin's 'Submerged'. *Anthropocenes – Human, Inhuman, Posthuman 4*(1), 4.

Burrows Edwin, G. and Wallace, Mike. 1999. *Gotham: A history of New York City to 1898*. New York: Oxford University Press.

Crutzen Paul J. 2002. Geology of mankind. *Nature 415*, 23.

Crutzen, Paul J. and Stoermer, Eugene F. 2000. The "anthropocene". *IGBP Newsletter 41* (Stockholm: Royal Swedish Academy of Sciences).

Dhillon, Jaskiran. 2017. *Prairie rising: Indigenous youth, decolonization, and the politics of intervention*. Toronto: University of Toronto Press.

Emejulu, Awkwugo. 2022. *Fugitive Feminism*. London: Silver Press.

Gallant, Y. 2023. *Israeli Defence Minister orders complete siege on Gaza* [Video] Al Jazeera. https://www.aljazeera.com/program/newsfeed/2023/10/9/israeli-defence-minister-orders-complete-siege-on-gaza

Goeman, Mishuana R. 2017. Ongoing storms and struggles: Gendered violence and resource exploitation. In Joanne Barker (ed.) *Critically sovereign: Indigenous gender, sexuality, and feminist studies* (pp. 99–126). Durham: Duke University Press.

Gómez-Barris, Macarena. 2017. *The Extractive Zone: Social ecologies and decolonial perspectives.* Durham: Duke University Press.

Guerin, Ayasha. 2019. *Submerged* [Video]. Unpublished.

Gumbs, Alexis Pauline. 2020. *Undrowned: Black feminist lessons from marine mammals.* Stirling: AK Press.

Hayward, Philip. 2012. Aquapelagos and aquapelagic assemblages. *Shima* 6(1), 1–11.

Hayward, Philip. 2015. The aquapelago and the estuarine city: Reflections on Manhattan. *Urban Island Studies 1*, 81–95.

Hewitt, John H. 1993. Mr. Downing and his Oyster house: The life and good works of an African-American entrepreneur – 19th Century New York, New York restaurateur, Thomas Downing. *New York History* 74(3), 229–252.

Holland, Sharon Patricia. 2023. *An other: A Black feminist consideration of animal life.* Durham: Duke University Press.

Hume, T. 2012. Young brothers, 'denied refuge,' swept to death by Sandy.' *CNN*, November 4. http:www.cnn.com/2012/11/02/world/americas/sandy-staten-island-brothers/

Hurley, Andrew. 1994. Creating ecological wastelands. *Journal of Urban History* 20(3), 340–364.

Jacques, Peter J. 2017. The origins of coastal ecological decline and the great Atlantic oyster collapse. *Political Geography 60*, 154–164.

Lewis, Simon L. and Maslin Mark A. 2015. Defining the anthropocene. *Nature* 519(7542), 177–180.

Luciano Dana. 2015. The inhuman anthropocene. *AVIDLY: LA Review of Books.* http://avidly.lareviewofbooks.org/2015/03/22/the-inhuman-anthropocene/

Mitchell, J. 1956. Mr. Hunter's grave. *The New Yorker.* www.newyorker.com/magazine/1956/09/22/mr-hunters-grave

Moore, Jason, M. 2016. *Anthropocene or capitalocene? Nature, history, and the crisis of capitalism.* Binghamton: PM Press.

Moten, Fred. 2003. *In the break: The aesthetics of the Black radical tradition.* Minneapolis: University of Minnesota Press.

Pritchard, Evan T. 2007. *Native New Yorkers: The legacy of the Algonquin people of New York.* San Francisco: Council Oak Books.

Reilly, Kevin S. 1993. Slavers in disguise: American whalemen and the Slave Trade 1845–1862. *The American Neptune* 53(3), 177–189.

Sanderson, Eric W. 2009. *Mannahatta: A natural history of New York city.* New York: Abrams.

Sharpe, Christina. 2016. *In the Wake: On Blackness and being.* Durham: Duke University Press.

Spillers, Hortense. J. 1987. Mama's baby, Papa's maybe: An American Grammar book. *Diacritics* 17(2), 64–81.

Strong, John A. 2018. *America's early whalemen, Indian Shore whalers on Long Island, 1650–1750.* Tuscon: University of Arizona Press.

Tallbear Kim. 2019. Caretaking relations, not American dreaming. *Kalfou* 6(1), 24–41.

Taylor Sanaura. 2017. *Beasts of burden: Animal and disability liberation.* New York: The New Press.

Todd, Zoe. 2017. Fish, kin and hope: Tending to water violations inamiskwaciwâskahikan and Treaty Six territory. *Afterall: A Journal of Art, Context and Enquiry 43*, 102–107.

Tsing, Anna Lowenhaupt. 2015. *The mushroom at the end of the world*. Princeton: Princeton University Press.

Françoise Vergès. 2017. Racial Capitalocene: Is the Anthropocene racial? Verso Blog post. https://www.versobooks.com/en-gb/blogs/news/3376-racial-capitalo cene?srsltid=AfmBOoqayBFRaLkA-KZCjAM9VnaL2IJ-zStMmUUajQ_HydS 5EdTN6F0g

Warsh Molly, A. 2010. Enslaved pearl divers in the sixteenth century Caribbean. *Slavery and Abolition* 31(3), 345–362.

Wynter, Sylvia. 2001. Towards the sociogenic principle: Fanon, identity, the puzzle of conscious experience, and what it is like to be "Black'. In Mercedes F. Durán-Cogan & Antonio Gómez-Moriana (eds.) *National identities and sociopolitical changes in Latin America* (1st ed., pp. 30-66). New York: Routledge.

Wynter, Sylvia. 1994. "No Humans Involved": An open letter to my colleagues. *Forum N.H.I.: Knowledge for the 21st Century* 1(1), 42–73.

Yazzie, Melanie K. and Baldy, Cutcha Rising. 2018. Introduction: Indigenous peoples and the politics of water. *Decolonization* 7(1), 1–18.

AFTERWORD

Things, Things That Matter and the Value of Aquapelagic Thinking

Mike Evans

Assemblage theory in contemporary humanities and social science research has been tremendously productive over the last two decades, no less in Island Studies than elsewhere. Indeed, one might argue (and I think I would), that the insistence on taking non-human entities seriously, systematically and as they manifest in the various entities/configurations that they may have, has been particularly important to Island Studies scholarship because of the sometimes naïve empiricism that assemblage theory has displaced. It might not be too dramatic to observe that assemblage theory-based approaches have precipitated an almost existential crisis. What is Island Studies as an area of research if islands as such (i.e. as ideal geological forms) no longer provide the boundary conditions for the field? Nonetheless, here we are, with the ontological basis of Island Studies fundamentally at risk. As Hayward notes (2025a), these challenges are a long time coming, with Epeli Hau'ofa's influential 1993 'Our Sea of Islands" post-colonial critique of the unexamined assumptions of colonial geographies as foundational in the reformation of Pacific Studies as Donna Haraway's (1988) "Situated Knowledges" (which effectively critiqued the idea of objectivity) was in the social sciences more generally. While neither scholar eschews an empirical path, both demand reflection on the positions from which such paths are navigated; and they reject the use of essentialized categories like "Island" as if the category is given in nature rather than knowledge systems themselves.

With the rise of assemblage theory (and assemblage and systems thinking), contemporary researchers have new tools to give both of these critiques shape. I will not rearticulate the roots of assemblage theory here (Hayward does an excellent job of that in the Introduction to this volume [2025a]), but

DOI: 10.4324/9781003569534-11

I do want to draw out some of the ontological implications of both Hau'ofa's and Hayward's work, and of assemblage theory in Island Studies more generally. The contributors to this volume provide examples, and pretty pervasive ones, as to why (from a heuristic perspective) the more marine based orientation of the "aquapelagic" presents productive new avenues of enquiry[1]. An assemblage, however complex it is internally, has a "thingness" about it, i.e. an assemblage is an ontological unit in the same way that an island is; one question about either is how useful they are, and to whom. An essentialized geography is a tool that relies on a Western analytical tradition that admits complex epistemologies but assumes a single (value-free) ontology; the subversion of that geography implicit in assemblage theory is thus ontological, with the very objects of investigation subject to variation and change (i.e. in evolving relational configurations). Not coincidentally, there are de-colonial consequences here as well.

Re-quoting Hau'ofa from Hayward's Introduction is warranted here – "Oceania is us. We are the sea, we are the ocean, we must wake up to this ancient truth and together use it to overturn all hegemonic views that aim ultimately to confine us". It is not accidental that the examples that Hayward draws on for his explication of the greater utility of the aquapelagic lens are drawn from the assemblages mobilized by Indigenous peoples; like Hau'ofa's Oceania, the aquapelagos of the Torres Straits and Haida Gwaii described by Hayward (2025a) are held together by the knowledge systems of Indigenous peoples as they create, re-create and transform assemblages through relational ontologies that demand wholistic thinking about the thing. In the language of current Indigenist research, de-colonization and Indigenization come together as Indigenous peoples struggle to protect their lands/waters. These assemblages are collective and fostered by ways of thinking and acting, including what the Indigenous methodologist Shawn Wilson calls relational accountability. Wilson argues that "as a researcher you are answering to all your relations when you are doing research" (2001,177) – such relations are multi-species and (effectively) actant rich. That such accountability is an attribute characteristic of Indigenous knowledge systems is well exemplified by Hayward's analysis of Hau'ofa; the examples he mobilizes to establish the more effective description of the not just vibrant matter, but how people, quite literally, relate. Again, the challenge here is not simply epistemological or even methodological, it is ontological, with Indigenous ontologies juxtaposed to the Western traditions that underpin much of contemporary science. Put another way, I am claiming here that Indigenous ontologies are always relational (and thus relative and flexible) and not coincidentally place-based rather than transcendent and universal (or universalizing).

The alternative to the traditions of Western analytical philosophy resident in Indigenous approaches is fundamentally ontological. Where European philosophical approaches assume that there is one universally applicable ontology (there may be many epistemologies but of ontologies there can be only one), Indigenous ontologies are understood as relative/relational, changing, and worthy of explication. This ontological challenge is especially useful in the context of climate change and anti-colonial transformations; and it helps us to reorientate things. The resonance with assemblage theory is obvious, and animation of the non-human other is assumed (in various ways and to various extents) in both as well. The case studies included in this volume provide rich examples of both how and why such work matters. I suppose that one of the comforting things about islands is that they have boundaries, but it is not at all obvious that such boundaries are helpful in the current context. The assembled articles in this volume demonstrate how the notion of aquapelagos, and, more importantly, aquapelagic thinking might be more useful, more helpful and more accountable to all our relations going forward. That the notion of an aquapelagic approach or the set of possible aquapelagic assemblages or the aquapelago itself can be reified is ontological peril best avoided; given the essential fluidity of water this is relatively easily done.

Jun'ichiro Suwa's chapter "Shima, Shimaguni and Aquapelagic Assemblages" (2025) is an excellent example of accountability in aquapelagic thinking. His notes that his earlier work on the "Japanese concept of *shima* reflects the idea of aquapelago well in that *shima* is an assemblage of geographic characteristics of islands with their waters and is inseparable from human activity", while *shimaguni* references a wider assemblage founded in politics and identity at a larger scale. Neither of these assemblages implies either boundedness or a direct correspondence to underlying (read essentialized) geographies as such, but they are profoundly place-based at the same time. He writes that these:

> aquapelagic assemblages … are not necessarily self-contained, self-sufficient and/or self-sustainable. Aquapelagic assemblages are specific products of ongoing processes in actual locations; and the resource uses, finances, identities and alliances involved in constituting them are fluid, transient and sometimes elusive.
>
> *(2025, xx)*

In other words, according to Suwa in an Indigenist view of Japan there are two forms of aquapelagic assemblages nested and interacting in place, and producing a very particular geo-politic, one that includes collective and relational values in the ideas of sanctuary. This does not mean that *shima* or *shimaguni* are products of human imagination. Consistent with the work of early assemblage theorists like Latour, they are complex and contingent but also self-replicating or iterative to some degree.

Aquapelagic thinking also helps bring more complex geographies into focus. Indeed, this volume does a particularly good job of holding reefs in relief, and several authors demonstrate the utility of the approach. For example, Fleury and Johnson (2025) analyze the eco-politics of the Écréhous and Minquiers reefs of Jersey, Chatterjee (2025) investigates the fisheries in the context of boundary disputes around the island of Katchatheevu and Vandenberg (2025) examines the impacts of attempts to recover/restore coral reefs in the Spermonde Archipelago. In the Écréhous and Minquiers, the geopolitical history of the region can be understood not only in the complex border politics of the English Channel and, indeed, the relations between Jersey and English/French dynastic history, but also in the materiality of the reefs, the fish and the fishers in relation to both. Similar dynamics are described by Chatterjee in the aquapelagic spaces centred on Katchatheevu, an island on the marine boundary between India and Sri Lanka subject to the geopolitics between the two but mediated by the practices of the fish and fishers of both countries. In both these examples, maps of terrestrial forms and national boundaries and treaties and disputes between nations give way to the assemblages of fish and fishers, both necessarily aquapelagic.

Vandenberg's (2025) Indonesian exemplar of a clash between the thinking of Indigenous Spermonde islanders and possibly well-intentioned interventions by conservationists to protect one particular coral reef area in the aquapelago demonstrates why (and how) aquapelagic thinking matters. In this example, conservationist attempts to protect, and clashing understandings of scale (one reef area versus the larger aquapelagic assemblage inhabited by the Indigenous residents), lead to conflict. The resistance of the local community is based on the accurate apprehension that the forces of restoration neither understand nor value the underlying Indigenous assemblage of which the reef is a part. This misunderstanding thus replicates a wider colonial assemblage which local people understand all too well. An appreciation for an aquapelagic lens (one hopes a local one), can ameliorate such tensions. Vanderburg concludes, that "at the very least, CRR initiatives in archipelagic regions must adopt an aquapelagic framework that recognises local and regional history, inter-island network systems and other social and cultural practices in order to move beyond the narrowly assumed benefits of coral restoration to equally assumed isolated island communities" (2025, xx).

Chant and Chacana's (2025) description of conservation efforts in the Juan Fernández Islands resonates with the Spermonde Archipelago case study. Here the intensely globalized exploitation of the islands through cruise ship visits were based, ironically, on the marketing of the islands as exotic and isolated. Efforts to mitigate over-exploitation of resources to meet the consumptive desires of international tourists were patterned by expressions of nationalism in the form of the creation of national parks, with very little in the way of effective protections. Only recently have place-based initiatives emerged, with an aquapelagic assemblage (as opposed to globalized ones) gaining ascendency. Here again descriptions of the interplay of colonial assemblages (including images of island isolation) and the struggle of Indigenous communities are managed through an aquapelagic lens that erodes essentialized islands while insisting on wider and more complex systems thinking. The authors here are actively linking islands and marine ecologies into a place-based (i.e. an aquapelagic) assemblage and opposing (not ignoring) such constructions to the colonial and national assemblages that take the islands as simple (even archipelagic) objects.

In the chapter "Sanctuary islands in a hostile matrix" Hayward's (2025c) examination of underwater salt diapirs (domes) in the Gulf of Mexico is instructive when thinking about our capacity for (and the consequences of) apprehending an aquapelagic assemblage. Fishing (and fisheries management) are notoriously prone to radical and sometimes destructive events based, in part, on changes in the ability of fishers to know aquapelagic conditions directly. Fishers look for signs, anticipate patterns and rely on complex knowledge systems to predict (however imperfectly) the location of fish and the health of fish stocks. New technological developments in fish detection enhance the capacity to know and catch fish, and they inadvertently devastate stocks. The northeastern North American cod fishery is a case in point; bigger trawlers used better sonar and radar to see and track what they could now catch and hold, without any corresponding mechanisms to limit those catches and plenty of political and economic pressures to ignore the warnings of inland fishers detecting an impending catastrophic collapse. However previous fishers understood the FGBs prior to the discovery of oil, the new prominence of the seabed led to the creation of a marine protection area and, with it, a new kind of legibility. The concept/figure of the aquapelago is equally apposite in terms of the FGBs' comprehension and representation by human agencies and to the subsequent eco- and geo-political discourses, strategies and legislative measures passed to restrict human use of its subsurface spaces, since it emphasizes the FGBs and NMSs as entities *performed by* an intersection of animate and inanimate objects and the abstract discourses applied to them (2025c, xx). Also instructive here is Hayward's treatment of fog on the Grand Banks; human technologies matter but so too do atmospheric conditions that obscure our capacity to apprehend. Indeed, fog banks are highly mobile and quite resistant to human efforts (pre-radar anyway) to penetrate (Hayward, 2025b). Aquapelagic assemblages not only include actants people learn to read, but they also include ones that confound and confuse.

Among the most hopeful applications of the aquapelagic lens in this volume are the two chapters focused on the Hudson River estuary – aka greater New York. Hayward and Joseph (2025) recount the development and loss of an Indigenous and then early colonial adaptations to the region that were based in practices responsive to ecological systems and based in relationality to more industrial and insulated practices that effectively ignored (and almost incidentally damaged) those assemblages. The capacity of the city to draw resources from globalized flows of wealth rendered local assemblages moot until Hurricane Sandy announced an era of renewed relevance to wider ecological processes. They examine recent attempts to imagine a new future for the city in which its place in the aquapelagic context of, and relationships with, the estuary are proactively reengaged. Readaption of the coastline and the adoption of more resilient land and waterscape features figure prominently here; and they represent reconstruction of relations with the wider assemblage in which the city is, in fact, located (i.e., the fact of rising water levels, more extreme weather systems, storm surges and climate change). These are potential exemplars for other great cities, not coincidentally located in similar aquapelagic contexts (i.e. where a river meets the sea, and people connect through both). What is not represented in these new New York imaginaries is the palpable culpability of the city in the wider contexts of the global processes now buffeting its shores. The least generous view here is that while the reengagement with complex local and regional assemblages is helpful, it is also an apologetic that falls short given the import of what happens in that estuary/bay.

Guerin's intervention (2025), both in their artistic practice and the histories mobilized in their chapter, are instructive in that the ecological/relational dynamics they attend to are attentive to dynamics of power in its various forms, mobilizing a concern for "Black and Indigenous ontologies" and the potential released by reciprocal relations with other species in aquatic spaces. In their most potent example, Guerin describes what I would characterize as the Indigenization of a Black community through its development of relational practices in the context of oyster beds off Staten Island. This community not only harvested from the beds, but it also proactively sought to protect and establish regenerative practices as part of these relations. Most tellingly, this community's adaptation to the estuary was developed in an aquapelagic space free (or maybe at least somewhat insulated) from racialized labor markets, but subsequently disrupted as the beds themselves were enclosed by new tenure arrangement, captured by industrial capital and destroyed by over-mechanization. Juxtaposed with the attempts to imagine post-Hurricane Sandy futures described by Hayward and Joseph (2025), the case reminds us that, however creative/responsive our new relations, these continue to exist in the context of the power, wealth and economic systems that displace local relations almost incidentally.

Guerin asks, "might attention extended to the survival of the non-human other bring emergent strategies for future survival?" (2025, xx). A fair question indeed. An answer has to be "yes" and insisting on the open and relational ontologies implicit in the development of aquapelagic thinking and explicit in Indigenous ontologies and methodologies is a good start.

Collectively the papers in this volume provide some much-needed examples of what relational geographies look like and are good for.

Note

1 I am going to resist the temptation to engage in a critical comparison between aquapelagic and archipelagic approaches to assemblage theory in Island Studies. Either is more useful than essentialized insular geography, and both open up new ways to construct assemblages/systems and explicate their dynamics.

References

Chant, Elizabeth and Chacana, Natalia Gandara. 2024. The Juan Fernandez Islands in transition. In Hayward, Philip and Joseph, May (Eds.) *Aquapelagos: Integrated Terrestrial and Marine Assemblages*. New Dehli and New York: Routledge, xx–xx.

Chatterjee, Arup. 2025. The Entangled Island: Katchatheevu and Indo-Lankan maritime relations. In Hayward, Philip and Joseph, May (Eds.) *Aquapelagos: Integrated Terrestrial and Marine Assemblages*. New Dehli and New York: Routledge, xx–xx.

Fleury, Christian and Johnson, Henry. 2025. Making aquapelagic place in Jersey: The Minquiers and Écréhous reefs. In Hayward, Philip and Joseph, May (Eds.) *Aquapelagos: Integrated Terrestrial and Marine Assemblages*. New Dehli and New York: Routledge, xx–xx.

Guerin, Ayasha. 2025. We, the submerged: (Non)Humans, race and aquapelagic relations – Notes from New York. In Hayward, Philip and Joseph, May (Eds.) *Aquapelagos: Integrated Terrestrial and Marine Assemblages*. New Dehli and New York: Routledge, xx–xx.

Haraway, Donna. 1988. Situated Knowledges: The Science Question in Feminism and the Privilege of Partial Perspective. *Feminist Studies*, 14(3), 575–599.

Hau'ofa, Epeli. 1993. Our sea of islands. In Waddell, Eric, Naidu, Vijay and Hau'ofa, Epeli (Eds.) *A New Oceania: Rediscovering our Sea of Islands*. Suva: University of the South Pacific, 2–16.

Hayward, Philip. 2025a. Introduction: The aquapelago as an integrated assemblage. In Hayward, Philip and Joseph, May (Eds.) *Aquapelagos: Integrated Terrestrial and Marine Assemblages*. New Dehli and New York: Routledge, xx–xx.

Hayward, Philip. 2025b. The precarious aquapelagic assemblage of the Grand Banks (northwest Atlantic). In Hayward, Philip and Joseph, May (Eds.) *Aquapelagos: Integrated Terrestrial and Marine Assemblages*. New Dehli and New York: Routledge, xx–xx.

Hayward, Philip. 2025c. The Flower Garden Banks and the parameters of aquapelagic sanctuary. In Hayward, Philip and Joseph, May (Eds.) *Aquapelagos: Integrated Terrestrial and Marine Assemblages*. New Dehli and New York: Routledge, xx–xx.

Hayward, Philip and Joseph, May. 2025. New York: Lenapehoking/New York: An estuarine aquapelago. In Hayward, Philip and Joseph, May (Eds.) *Aquapelagos: Integrated Terrestrial and Marine Assemblages*. New Dehli and New York: Routledge, xx–xx.

Suwa, Juni. 2025. *Shima, shimaguni* and aquapelagic assemblages. In Hayward, Philip and Joseph, May (Eds.) *Aquapelagos: Integrated Terrestrial and Marine Assemblages*. New Dehli and New York: Routledge, xx–xx.

Vandenberg, Jessica. 2025. Colonial legacies and restoration futures: Examining the risks of dispossession from coral reef restoration in the Indonesian aquapelago. In Hayward, Philip and Joseph, May (Eds.) *Aquapelagos: Integrated Terrestrial and Marine Assemblages*. New Dehli and New York: Routledge, xx–xx.

Wilson, Shawn. 2001. What is an Indigenous research methodology? *Canadian Journal of Native Education*, 175–179.

INDEX

Sethusamudram Shipping Canal project
122, 128–129
Shagapova, V. S. 53
Shastri, Lal Bahadur 126
Shihei, Hayashi 24
shima 2–3, 22, 25–29, 84, 89, 92, 175
Shimaguni 23–25, 175
Shima-okuri/shima-nagashi 24
Shimaumi ("island-bearing") 23
Shima-watari ("came across the islands") 29
Shunkan 24
Sinhalese fishermen 120
Sinhalese Lankan navy 120
Sinhalese majoritarianism 126
"Situated Knowledges" 173
Skottsberg, Carl 109, 110
slave ships 165
"socio-oceanography" 3
South Moresby Agreement 15
Spermonde Archipelago xxi, 65, 66, 68, 176
Spillers, Hortense. J. 164
Springer, Anne Sophie 95
Sri Lanka's Regulation of Foreign Fishing Boats Act 132
Staten Island 146
Steinberg, Ted xix, xxii
Stoermer, Eugene F. 157
St Pierre and Miquelon 50
Strong, John A. 164
Subercaseaux, Benjamín 106
Sunderban Islands 134
Suwa, Jun'ichiro 89–90, 101, 175

Tagliacozzo, Eric xix
Tagore, Rabindranath 133
tairiku-ronin 24
Tallbear Kim 161
Tamil Eelam naval wing ("Sea Tigers") 128
Taylor, Sunaura 161, 165
Teach me how fer lew (King) 11, 12
terra nullius 10
terraqueous boundaries xix
Territorial Sea Act 43
Tiaoyutai Islands 123
Title III of the US Marine Protection, Research, and Sanctuaries Act 89
Todd, Zoe 166
Tomoya, Akimichi 27
Torres Strait Islands 8–11, *9*

Toshio, Shimao 25
Tourism displacement theories 77
Trade and Cooperation Agreement (TCA) 39
transatlantic slave trade 162
"trans-corporeality" framework 159
Treasure Island (Stevenson) 105
The Treasury of Newfoundland Stories (James) 57
Treaty of Peace and Friendship 123
Trump, Donald 92, 96
Tsing, Anna Lowenhaupt 22, 159
Tsuneichi, Miyamoto 25
Turpin, Etienne 95
Turtle Island 140, 154

Undrowned (Gumbs) 166
United Nations Convention on the Law of the Sea 124
US Department of Homeland Security Navigation Center of Excellence (USDHSNCE) 48
US Marine Protection, Research, and Sanctuaries Act 91
US National Marine Sanctuary System 89
utaki 27
uti possidetis juris (ptinciple) 127

Vale, Celina 52
Valmiki 122
Vergès, Françoise 158
Vibrant Matter (Bennett) xix, 159–160
Vienna Convention on the Law of Treaties 133
Villiers, Alan, J. 59, 60
Vulnerable Marine Ecosystem (VME) 113 see also Juan Fernández Islands
Vypin Island 134

Ward Island 146
Watt, James 157
We, the Submerged (Guerin) 162–164, 166
"weather bombs" 49
West, Paige 67
Whaling in the Frozen South (Villiers) 59
whaling ships 165
Whittaker, Robert J. 84
Wigen, Rebecca J. 14